SAUSSURE

Jonathan Culler

THE HARVESTER PRESS
By agreement with Fontana

THE HARVESTER PRESS LIMITED
Publisher: John Spiers
2 Stanford Terrace,
Hassocks, Sussex

Saussure
This edition first published in 1976 by
The Harvester Press Limited
by agreement with Fontana.
Published simultaneously in paperback
in Great Britain by Fontana Books in
their 'Modern Masters' series edited by
Frank Kermode.

Copyright © 1976 Jonathan Culler

ISBN 0 85227 379 8

Printed in Great Britain by
Biddles Limited, Guildford, Surrey

Contents

In Memoriam
Veronica Forrest-Thomson
1947-1975

Introduction

Ferdinand de Saussure is the father of modern linguistics, the man who re-organized the systematic study of language and languages in such a way as to make possible the achievements of twentieth-century linguistics. This alone would make him a Modern Master: master of a discipline which he made modern. But he has other claims to our attention as well.

First of all, together with his two great contemporaries, Emile Durkheim in sociology and Sigmund Freud in psychology, he helped to set the study of human behaviour on a new footing. These three men realized that one could not approach an adequate understanding of man and his institutions if one treated human behaviour as a series of events similar to events in the physical world. A scientist can study the behaviour of objects under certain conditions, such as the trajectories of projectiles fired at different angles and velocities, or the reactions of a chemical substance to a variety of temperatures. He can describe what happens and try to explain why without paying any attention to ordinary people's impressions or ideas about these matters. But human behaviour is different. When studying human behaviour the investigator cannot simply dismiss as subjective impressions the meaning behaviour has for members of a society. If people see certain actions as impolite, that is a fact which directly concerns him, a social fact. To ignore the meanings which actions and objects have in a society would be to study mere physical events. Anyone analysing human behaviour is concerned not with events themselves but with events that have meaning.

Moreover, Saussure, Freud, and Durkheim saw that the study of human behaviour misses its best opportunities if it tries to trace the historical causes of individual events.

7

Instead it must focus primarily on the functions which events have within a general social framework. It must treat social facts as part of a system of conventions and values. What are the values and conventions which enable men to live in society, to communicate with one another, and generally to behve as they do? If one tries to answer these questions the result is a discipline very different from that which replies to questions about the historical causes of various events. Saussure and his two contemporaries established the supremacy of this type of investigation, which looks for an underlying system rather than individual causes, and they thus made possible a fuller and more apposite study of man.

Secondly, by his methodological example and by various prophetic suggestions which he offered, Saussure helped to promote semiology, the general science of signs and systems of signs, and structuralism, which has been an important trend in contemporary anthropology and literary criticism as well as in linguistics. Indeed, the revival of interest in Saussure in the past few years is largely due to the fact that he has been the inspiration for semiology and structuralism as well as for structural linguistics.

Thirdly, in his methodological remarks and general approach to language Saussure gives us a clear expression of what we might call the formal strategies of Modernist thought: the ways in which scientists, philosophers, artists, and writers working in the early part of this century tried to come to terms with a complex and chaotic universe. How does one cope, systematically, with the apparent chaos of the modern world? This question was being asked in a variety of fields, and the replies which Saussure gives – that you cannot hope to attain an absolute or God-like view of things but must choose a perspective and that within this perspective objects are defined by their relations with one another rather than by essences of some kind – are exemplary. Saussure enables us to grasp with unusual clarity the strategies of Modernist thought.

Finally, Saussure's treatment of language focuses on

8

problems which are central to new ways of thinking about man and especially about the intimate relation between language and the human mind. If man is indeed the 'language animal', a creature whose dealings with the world are characterized by the structuring and differentiating operations which are most clearly manifested in human language, then it is Saussure who set us on his track. When we speak of the human tendency to organize things into systems by which meaning can be transmitted, we place ourselves in what is very much a Saussurian line of thought.

These contributions – to linguistics, to the social sciences generally, to semiology and structuralism, to Modernist thought and to our conception of man – make Saussure a seminal figure in modern intellectual history. This book, therefore, must range widely over linguistics, semiology, philosophy, and the social sciences if it is to define Saussure's importance. But paradoxically, Saussure himself wrote nothing of general significance. A book on the vowel system of early Indo-European language, a doctoral thesis on the use of the genitive case in Sanskrit, and a handful of technical papers are all that he ever published. Nor did he leave behind a rich hoard of unpublished writings. His influence, both within and beyond linguistics, is based on something he never wrote. Between 1907 and 1911, as Professor at the University of Geneva, he gave three courses of lectures on general linguistics. After his death in 1913 his students and colleagues decided that his teachings should not be lost and constructed out of various sets of lecture notes a volume entitled *Cours de linguistique générale*, a course in general linguistics.

We shall have more to say in Chapter One about the strange genesis of the *Course*, the way in which the published text was put together. For the moment the important point is this: whatever Saussure's general significance for modern thought – and it is considerable – he himself was first and foremost, perhaps even exclusively, a linguist, a student of language. Someone who knows Saussure only by reputation, as founder of modern linguistics, promoter of a new conception of language, and inspiration for anthropologists and literary critics, might expect to find the *Course in General Linguistics* a book full of broad generalizations, portentous observations about the nature of language and mind, elaborate and eloquent theories about man as a social and communicative being. In fact, nothing could be further from the truth. What strikes one most forcibly in the *Course* is Saussure's active and scrupulous concern for the foundations of his subject.

His concern with the nature of language and the foundations of linguistics takes the form of a questioning of the assumptions we make when we talk about language. For example, if you make a noise and at some other time I make a noise, under what conditions would we be justified in saying that we had uttered the same words? Such questions may seem trivial. One might be tempted to reject them as pointless quibbling, arguing that we simply *know* whether two people have uttered the same words or not. But the point is, how do we know? What is involved in knowing this? For whatever is involved is part of our knowledge of language, our knowledge of the units of that language. Such questions are far from trivial. If we are to analyse a language we must be able to form a clear and coherent idea of the units or elements of that language. If, for example, we are to think of the 'word' as a unit of language, then we must know how we determine that two people have uttered the same word, though the actual physical sounds they made were different.

Saussure asks fundamental and probing questions which linguists before him had failed to ask, and he provides answers which have revolutionized the way in which language is studied. Though the solutions and definitions he offers might initially seem of interest only to students of linguistics, they have direct bearing on the fundamental problems of what the French call the 'human sciences': the disciplines which deal with the world of meaningful objects and actions (as opposed to physical objects and events themselves). Saussure's reflections on the sign and on sign-systems pave the way for a general study of the ways in which human experience is organized.

This wider significance is doubtless of greater interest to readers of this book than debates about the precise nature of Saussure's distinctions and linguistic categories, and therefore discussion in the following chapters will always aim towards larger issues. But if we are to grasp the radical implications of Saussure's ideas we must follow the logic of his argument in some detail. We must go back, with Saussure, to first principles and ask elementary questions about human language, about the nature of the sign, about the identity of units of a language. We must begin by exploring Saussure's theory of language.

This is not an easy task. It requires detailed explication. That it is not an easy task is amply shown by the fact that Saussure himself did not feel in a position to write a course in general linguistics. If he had believed that he had solved the fundamental problems of linguistics in an unequivocal way, if he had not felt that he was still groping towards a satisfactory formulation of ideas which he but glimpsed, doubtless he would have written the book himself. Since he did not, we must make an effort to grasp a thought which is not yet fully born but which, even in its nascent state, was able to exert a powerful influence on succeeding generations of linguists.

Our first task, therefore, after a brief look at Saussure's life and the circumstances which led to the publication of the *Course*, is to explore Saussure's theory of language: to begin with first principles and to reconstruct the foundations of modern linguistics. Thus equipped, we can undertake the second task which is essential if we are to understand Saussure and the significance of his work. The *Course* arose from Saussure's dissatisfaction with the theoretical foundations of linguistics as then practised. What was the state of linguistics as Saussure saw it? How does his work fit into the history of linguistics, the history of thought about language? Then, in Chapter Four we can turn from the past to the present and future and outline the significance of

Introduction

Saussure's work for semiology, the general science of signs which he envisaged but which did not really begin to take shape until many years after his death.

Following the fortunes of Saussure's ideas in linguistics and semiology, tracing their actual influence, is doubtless our central task; but if we are to sum up his significance for twentieth-century thought we must also attempt to bring into the open those aspects of his work which, inadequately formulated in the *Course*, have often been misconstrued or ignored. In this way we may try to ensure that Saussure be considered not only an important figure of the recent past but also, and perhaps especially, a major intellectual presence today.

August 1975   *Brasenose College, Oxford*

1   The Man and the *Course*

Saussure is a fascinating and enigmatic figure because he lived such an uneventful life. As far as we can tell, he had no great intellectual crises, decisive moments of insight or conversion, or momentous personal adventures. His own modesty about his thought, bold and uncompromising though that thought was, makes it very difficult to trace its genesis in his earlier intellectual life, and the fact that his major work should have remained unwritten seems the appropriate climax to this paradoxical career.

Born in Geneva in 1857, one year after Freud and one year before Durkheim, Saussure was the son of an eminent naturalist and member of a family with a strong tradition of accomplishment in the natural sciences. He was introduced to linguistic studies at an early age by a philologist and family friend, Adolphe Pictet. At the age of fifteen, after he had learned Greek to add to his French, German, English, and Latin, Saussure tried to work out a 'general system of language' and wrote for Pictet an 'Essay on Languages' in which he argued that all languages have their root in a system of two or three basic consonants. Though Pictet must have smiled at the extreme reductionism of this youthful attempt, he did not discourage his protégé, who began to study Sanskrit while still at school.

In 1875 Saussure entered the University of Geneva but, following family tradition, enrolled as a student of physics and chemistry, though continuing to follow courses on Greek and Latin grammar. This experience convinced him that his career lay in the study of language, for not only did he join a professional linguistic association, the Linguistic Society of Paris, but, feeling that his first year at Geneva had been largely wasted, he persuaded his parents

13

to send him to the University of Leipzig to study Indo-European languages.

Leipzig was a fortunate choice: it was the centre for a school of young historical linguists, the *Junggrammatiker* or 'Neo-grammarians', and for the first time Saussure was able to match wits with the most creative linguists of his day. His sense of his own powers was doubtless confirmed when one of his Leipzig teachers, Brugmann, discovered what is called the law of *nasal sonans*, which Saussure had postulated several years earlier but rejected because it conflicted with the hypotheses of eminent linguists.

For four years Saussure remained in Leipzig, except for an interlude of eighteen months in Berlin, and in December 1878, when he was 21, published his *Mémoire sur le système primitif des voyelles dans les langues indo-européennes* (Memoir on the Primitive System of Vowels in Indo-European Languages), which one linguist has called 'the most splendid work of comparative philology ever written'. The argument and conclusions of this work will be discussed in Chapter Three, but what is most striking about it is that the young linguist should have attacked a large and fundamental problem in historical linguistics and should have emphasized the importance of methodological problems. 'I am not speculating', he wrote in his Preface, 'on abstruse theoretical matters but enquiring into the very basis of the subject, without which everything is unanchored, arbitrary, and uncertain.'

The *Mémoire* was well received in many quarters, and when Saussure returned to Leipzig from Berlin he was asked by one professor whether he was by any chance related to the great Swiss linguist, Saussure, the author of the *Mémoire*. However, Saussure seems to have found Germany uncongenial, and after defending his thesis on the use of the genitive case in Sanskrit (for which he was awarded his doctorate *summa cum laude*) he left for Paris.

In France he was a considerable success. Soon after his arrival he began to teach Sanskrit, Gothic, and Old High German at the École pratiques des hautes études and after

1887 expanded his teaching to cover Indo-European philology in general. He was active in the Société linguistique de Paris and a major formative influence on the younger generation of French linguists. But in 1891, when offered a Professorship at Geneva, he decided to return to Switzerland, and even the honour which his older colleagues did him in having him named Chevalier de la Légion d'Honneur could not hold him in Paris.

In Geneva his students were fewer and less advanced; he taught Sanskrit and historical linguistics generally. He married, fathered two sons, rarely travelled, and seemed to be settling into a decent provincial obscurity. He wrote less and less, and then painfully, reluctantly. In a letter of 1894, one of the few revealing personal documents we possess, he refers to an article which he has finally surrendered to an editor and continues,

... but I am fed up with all that, and with the general difficulty of writing even ten lines of good sense on linguistic matters. For a long time I have been above all preoccupied with the logical classification of linguistic facts and with the classification of the points of view from which we treat them; and I am more and more aware of the immense amount of work that would be required to show the linguist *what he is doing* ... The utter inadequacy of current terminology, the need to reform it and, in order to do that, to demonstrate what sort of object language is, continually spoils my pleasure in philology, though I have no dearer wish than not to be made to think about the nature of language in general. This will lead, against my will, to a book in which I shall explain, without enthusiasm or passion, why there is not a single term used in linguistics which has any meaning for me. Only after this, I confess, will I be able to take up my work at the point I left off.[1]

He never wrote the book. He worked on Lithuanian, on medieval German legends, on a theory that Latin poets had concealed anagrams of proper names in their verses. But in 1906, on the retirement of another professor, the University assigned him responsibility for the teaching of general linguistics, and thenceforth, in alternate years (1907,

1908-9, 1910-11), he gave the lectures which were ultimately to become the *Cours de linguistique générale*. In the summer of 1912 he fell ill and died in February 1913 at the age of 56.

Saussure's career, though highly successful, was in no way extraordinary. His published writings would have assured him an honourable place in the history of philology, but a place roughly equivalent to that of other eminent Neo-grammarians such as Brugmann and Verner, who are today known only to philologists. Fortunately, Saussure's students and colleagues thought that his work in general linguistics should be preserved and produced the volume which makes him a seminal thinker.

It was not an easy task. As Bally and Sechehaye recount in their preface to the *Course*, Saussure had kept very few notes, so they had to work from notes taken by students who had attended the various series of lectures. But even when, from collation and comparison of notes, one gained a fair idea of what was said in each of the three lecture series, a major problem remained. To publish rough transcripts of all three series would involve enormous repetition (not to speak of inconsistencies), but to publish only one series would be to omit a good deal, since Saussure seemed to have composed each course afresh according to a different plan. Faced with this problem, Bally and Sechehaye, colleagues who had not themselves attended the lectures, made a bold decision which has been largely responsible for Saussure's influence. They decided to compose a unified work, to attempt a synthesis, granting precedence to the third series of lectures but drawing heavily on material from the other two and on Saussure's personal notes.

Most teachers would shudder at the thought of having their views handed on in this way, and it is indeed extraordinary that this unpromising procedure, fraught with possibilities of misunderstanding and compromise, should have produced a major work. But the fact is there: the *Course in General Linguistics*, as created by Bally and

Sechehaye, is the source of Saussure's influence and reputation. Not until 1967, when Rudolf Engler began to publish
the students' notes from which the *Course* was constructed,
was it possible to go very far beyond the constructed text.
It was the *Course* itself which influenced succeeding generations of linguists.

This fact poses something of a problem for our discussion.
On the one hand, Saussure's importance in linguistics and in
other fields rests less on what he 'really' thought than on
what is contained in the *Course*. On the other hand, the
availability of the students' notes makes one wish to point
out where the editors seem to have taken liberties, mis-
understood, or falsified Saussure's thought. In general they
did an admirable job, but there is a strong case for saying
that in three respects they were less successful than one
might have wished: their order of presentation is probably
not that which Saussure would have chosen and thus does
not reflect the potential logical sequence of his argument;
the notion of the arbitrary nature of the sign receives much
less discussion than it does in the notes; and in discussing
the sound plane of language the editors are much less
scrupulous and consistent in their terminology than
Saussure seems to have been. These are important matters
which one cannot wholly neglect, and thus in the discussion
that follows, though I shall be primarily concerned with the
Course itself, I shall occasionally attempt, especially through
the order of presentation, to reconstruct more exactly what
I take to be the logic of Saussure's thought. The major
emphasis falls on the Saussurian teachings of the *Course* and
their place in the history of linguistics, but in the exposition
of Saussure's theory of language, to which we now turn, I
shall not hesitate to rectify the original editors' occasional
lapses.

2 Saussure's Theory of Language

Saussure was unhappy with linguistics as he knew it because he thought that his predecessors had failed to think seriously or perceptively about what they were doing. Linguistics, he wrote,[1] 'never attempted to determine the nature of the object it was studying, and without this elementary operation a science cannot develop an appropriate method' (*Course*, 3; *Cours*, 16).

This operation is all the more necessary because human language is an extremely complex and heterogeneous phenomenon. Even a single speech act involves an extraordinary range of factors and could be considered from many different, even conflicting points of view. One could study the way sounds are produced by the mouth, vocal cords, and tongue; one could investigate the sound waves which are emitted and the way they affect the hearing mechanism. One could consider the signifying intention of the speaker, the aspects of the world to which his utterance refers, the immediate circumstances of the communicative context which might have led him to produce a particular series of noises. One might try to analyse the conventions which enable speaker and listeners to understand one another, working out the grammatical and semantic rules which they must have assimilated if they are to communicate in this way. Or again, one could trace the history of the language which makes available these particular forms at this time.

Confronted with all these phenomena and these different perspectives from which one might approach them, the linguist must ask himself what he is trying to describe. What in particular is he looking at? What is he looking for? What, in short, is language?

Saussure's answer to this question is unexceptionable but

extremely important, since it serves to direct attention to essentials. Language is a system of signs. Noises count as language only when they serve to express or communicate ideas; otherwise they are just noise. And to communicate ideas they must be part of a system of conventions, part of a system of signs. The sign is the union of a form which signifies, which Saussure calls the *signifiant* or signifier, and an idea signified, the *signifié* or signified. Though we may speak of signifier and signified as if they were separate entities, they exist only as components of the sign. The sign is the central fact of language, and therefore in trying to separate what is essential from what is secondary or incidental we must start from the nature of the sign itself.

THE ARBITRARY NATURE OF THE SIGN

The first principle of Saussure's theory of language concerns the essential quality of the sign. The linguistic sign is arbitrary. A particular combination of signifier and signified is an arbitrary entity. This is a central fact of language and linguistic method. 'No one', he writes,

contests the principle of the arbitrary nature of the sign, but it is often easier to discover a truth than to assign it its rightful place. The above principle dominates the whole of linguistic analysis of a language. Its consequences are innumerable, though they are not all, it is true, equally evident straight away. It is after many detours that one discovers them, and with them the fundamental importance of this principle (*Course*, 68; *Cours*, 100).

What does Saussure mean by the arbitrary nature of the sign? In one sense the answer is quite simple. There is no natural or inevitable link between the signifier and the signified. Since I speak English I may use the signifier represented by *dog* to talk about an animal of a particular species, but this sequence of sounds is no better suited to that purpose than another sequence. *Lod, tet,* or *bloop* would serve equally well if they were accepted by members of my speech community. There is no intrinsic reason why one of these signifiers rather than another should be linked with the concept of a 'dog'.*

Are there no exceptions to this basic principle? Certainly. There are two ways in which linguistic signs may be motivated, that is to say, made less arbitrary. First, there are cases of onomatopœia, where the sound of the signifier seems in some way mimetic or imitative, as in the English *bow-wow* or *arf-arf* (cf. French *ouâ-ouâ*, German *wau-wau*, Italian *bau-bau*). But there are few such cases, and the fact that we identify them as a separate class and special case only emphasizes more strongly that ordinary signs are arbitrary.

However, within a particular language signs may be partially motivated in a different way. The machine on which I am writing is called a *typewriter*. There is no intrinsic reason why it should not be called a *grue* or a *blimmel*, but within English *typewriter* is motivated because the meanings of the two sound sequences which compose its signifier, *type* and *writer*, are related to its signified, to the notion of a 'typewriter'. We might call this 'secondary motivation'. Notice, for example, that only in English is the relation between sound-sequence and concept motivated. If the French were to use the same form to speak of this machine, that would be a wholly arbitrary sign, since the primary constituent, *writer* is not a sign in the French language. Moreover, for Saussure, as we shall see later, the process of combining *type* and *writer* to create a new motivated sign is fundamentally similar to the way in which we combine words to form phrases (whose meaning is related to the combined meanings of individual words). We can say, therefore, that all languages have as their basic elements arbitrary signs. They then have various processes for combining these signs, but that does not alter the essential nature of language and its elementary constituents.

The sign is arbitrary in that there is no intrinsic link between signifier and signified. This is how Saussure's principle is usually interpreted, but in this form it is a

*Note that here, as throughout, I use italics to cite linguistic forms (e.g. *dog*, *lod*) and quotation marks to designate meanings (e.g. 'dog').

wholly traditional notion, a rather obvious fact about language. Interpreted in this limited way, it does not have the momentous consequences which, according to the students' notes, Saussure repeatedly claimed for it: 'the hierarchical place of this truth is at the very summit. It is only little by little that one recognizes how many different facts are but ramifications, hidden consequences of this truth' (Engler, 153). There is more to the arbitrary nature of the sign than the arbitrary relation between signifier and signified. We must push further.

From what I have said so far about signifier and signified, one might be tempted to think of language as a nomenclature: a series of names arbitrarily selected and attached to a set of objects or concepts. It is, Saussure says, all too easy to think of language as a set of names and to make the biblical story of Adam naming the beasts an account of the very nature of language. If one says that the concept 'dog' is rendered or expressed by *dog* in English, *chien* in French and *Hund* in German, one implies that each language has an arbitrary name for a concept which exists prior to and independently of any language.

If language were simply a nomenclature for a set of universal concepts, it would be easy to translate from one language to another. One would simply replace the French name for a concept with the English name. If language were like this the task of learning a new language would also be much easier than it is. But anyone who has attempted either of these tasks has acquired, alas, a vast amount of direct proof that languages are not nomenclatures, that the concepts or signifieds of one language may differ radically from those of another. The French 'aimer' does not go directly into English; one must choose between 'to like' and 'to love'. 'Démarrer' includes in a single idea the English signifieds of 'moving off' and 'accelerating'. English 'to know' covers the area of two French signifieds, 'connaître' and 'savoir'. The English concepts of a 'wicked' man or of a 'pet' have no real counterparts in French. Or again, what English calls 'light blue' and 'dark blue' and treats as two shades of a single colour are in Russian two distinct primary colours. Each language articulates or organizes the world differently. Languages do not simply name existing categories, they articulate their own.

Moreover, if language were a set of names applied to independently-existing concepts, then in the historical evolution of a language the concepts should remain stable. Signifiers could evolve; the particular sequence of sounds associated with a given concept might be modified; and a given sequence of sounds could even be attached to a different concept. Occasionally, of course, a new sign would have to be introduced for a new concept which had been produced by changes in the world. But the concepts themselves, as language-independent entities, would not be subject to linguistic evolution.

In fact, though, the history of languages is full of examples of concepts shifting, changing their boundaries. The English word *cattle*, for example, at one point meant property in general, then gradually came to be restricted to four-footed property only (a new category), and finally attained its modern sense of domesticated bovines. Or again, a 'silly' person was once happy, blessed, and pious. Gradually this particular concept altered; the old concept of 'silliness' transformed itself, and by the beginning of the sixteenth century a silly person was innocent, helpless, even deserving of pity. The alteration of the concept continued until eventually a silly person was simple, foolish, perhaps even stupid.

If language were a nomenclature we should be obliged to say that there exist a number of distinct concepts and that the signifier *silly* was attached first to one and then to another. But clearly this is not what happened: the concept attached to the signifier *silly* was continually shifting its boundaries, gradually changing its semantic shape, articulating the world in different ways from one period to the next. And, incidentally, the signifier also evolved, undergoing a modification of its central vowel.

What is the significance of this? What does it have to do with the arbitrary nature of the sign? Language is not a nomenclature and therefore its signifieds are not pre-existing concepts but changeable and contingent concepts which vary from one state of a language to another. And since the relation between signifier and signified is arbitrary, since there is no necessary reason for one concept rather than another to be attached to a given signifier, there is therefore no defining property which the concept must retain in order to count as the signified of that signifier. The signified associated with a signifier can take any form; there is no essential core of meaning that it must retain in order to count as the proper signified for that signifier. The fact that the relation between signifier and signified is arbitrary means, then, that since there are no fixed universal concepts or fixed universal signifiers, the signified itself is arbitrary, and so is the signifier. We then must ask, as Saussure does, what defines a signifier or a signified, and the answer leads us to a very important principle: both signifier and signified are purely relational or differential entities. Because they are arbitrary they are relational. This is a principle which requires explanation.

THE NATURE OF LINGUISTIC UNITS

Saussure attaches great importance – more so than it would appear from the published *Course* – to the fact that language is not a nomenclature, for unless we grasp this we cannot understand the full ramifications of the arbitrary nature of the sign. A language does not simply assign arbitrary names to a set of independently existing concepts. It sets up an arbitrary relation between signifiers of its own choosing on the one hand, and signifieds of its own choosing on the other. Not only does each language produce a different set of signifiers, articulating and dividing the continuum of sound in a distinctive way; each language produces a different set of signifieds; it has a distinctive and thus 'arbitrary' way of organizing the world into concepts or categories.

It is obvious that the sound sequences of *fleuve* and

rivière are signifiers of French but not of English, whereas *river* and *stream* are English but not French. Less obviously but more significantly, the organization of the conceptual plane is also different in English and French. The signified 'river' is opposed to 'stream' solely in terms of size, whereas a 'fleuve' differs from a 'rivière' not because it is necessarily larger but because it flows into the sea, while a 'rivière' does not. In short, 'fleuve' and 'rivière' are not signifieds or concepts of English. They represent a different articulation of the conceptual plane.

The fact that these two languages operate perfectly well with different conceptual articulations or distinctions indicates that these divisions are not natural, inevitable, or necessary, but, in an important sense, arbitrary. Obviously it is important that a language has ways of talking about flowing bodies of water, but it can make its conceptual distinctions in this area in any of a wide variety of ways (size, swiftness of flow, straightness or sinuosity, direction of flow, depth, navigability, etc.). Not only can a language arbitrarily choose its signifiers; it can divide up a spectrum of conceptual possibilities in any way it likes.

Moreover, and here we come to an important point, the fact that these concepts or signifieds are arbitrary divisions of a continuum means that they are not autonomous entities, each of which is defined by some kind of essence. They are members of a system and are defined by their relations to the other members of that system. If I am to explain to someone the meaning of *stream* I must tell him about the difference between a stream and a river, a stream and a rivulet, etc. And similarly, I cannot explain the French concept of a 'rivière' without describing the distinction between 'rivière' and 'fleuve' on the one hand and 'rivière' and 'ruisseau' on the other.

Colour terms are a particularly good example of this characteristic of the sign. Suppose we wish to teach a foreigner about colours in English. Let us suppose also that he is a rather slow learner from a non-European culture, so that we must work out an efficient teaching strategy. It might occur to us that the best way to proceed would be to take one colour at a time: to begin, for example, with brown and not go on to another colour until we were certain that he had mastered brown. So we begin by showing him brown objects and telling him that they are brown. Since we want to be thorough, we have assembled a collection of a hundred brown objects of various kinds. And then, after having bored him and ourselves for several hours, we take him into another room and, to test his knowledge of 'brown', ask him to pick out all the brown objects. He sets to work but seems to be having difficulty deciding what to select, so in despair we decide we haven't been thorough enough and propose to start again the next day with five hundred brown objects.

Fortunately, most of us would not adopt this desperate solution and would recognize what had gone wrong. However many brown objects we may show him, our pupil will not know the meaning of *brown*, and will not be able to pass our test, until we have taught him to distinguish between brown and red, brown and tan, brown and grey, brown and yellow, brown and black. It is only when he has grasped the relation between brown and other colours that he will begin to understand what brown is. And the reason for this is that brown is not an independent concept defined by some essential properties but one term in a system of colour terms, defined by its relations with the other terms which delimit it.

Indeed, this painful teaching experience would bring us to understand that because the sign is arbitrary, because it is the result of dividing a continuum in ways peculiar to the language to which it belongs, we cannot treat the sign as an autonomous entity but must see it as part of a system. It is not just that in order to know the meaning of *brown* one must understand *red*, *tan*, *grey*, *black*, etc. Rather, one could say that the signifieds of colour terms are nothing but the product or result of a system of distinctions. Each language, in dividing the spectrum and distinguishing categories which it calls colours, produces a different system of signifieds: units whose value depends on their relations with one another. As Saussure says, generalizing the point,

in all cases, then, we discover not *ideas* given in advance but *values* emanating from the system. When we say that these values correspond to concepts, it is understood that these concepts are purely differential, not positively defined by their content but negatively defined by their relations with other terms of the system. Their most precise characteristic is that they are what the others are not (*Course*, 117; *Cours*, 162).

Brown is what is not red, black, grey, yellow, etc., and the same holds for each of the other signifieds.

This is a major though paradoxical consequence of the arbitrary nature of the sign, and we shall return to it shortly. But perhaps the easiest way to grasp this notion of the purely relational nature of linguistic units is to approach it from another angle.

Consider the problem of identity in linguistics: the question of when two utterances or portions of an utterance count as examples of a single linguistic unit. Suppose someone tells me, 'I bought a bed today', and I reply, 'What sort of bed?' What do we mean when we say that the same sign has been employed twice in this brief conversation? What is the basis on which we can claim that two examples or instances of the same linguistic unit have appeared in our dialogue? Note that we have already begged the question in transcribing a portion of the noises that each of us made as *bed*. In fact, the actual noises which were produced will have been measurably different – different from a purely physical and acoustic point of view. Voices vary; after a very few words we can recognize a friend's voice on the telephone because the actual physical signals he emits are different from those of our other acquaintances.

My interlocutor and I produced different noises, yet we want to say that we have produced the same signifier, used the same sign. The signifier, then, is not the same thing as the noises that either he or I produced. It is an abstract unit of some kind, not to be confused with the actual sequence of sounds. But what sort of unit is it? Of what does the unit consist? We might approach this question by asking how far the actual noises produced could vary and still count as versions of the same signifier. This, of course, is similar to the question we implicitly asked earlier about the signified: how far can a colour vary and still count as brown? And the answer for the signifier is very similar to the answer for the signified. The noises made can vary considerably (there is no essential property which they must possess) so long as they do not become confused with those of contrasting signifiers. We have considerable latitude in the way we utter *bed*, so long as what we say is not confused with *bad*, *bud*, *bid*, *bode; bread*, *bled*, *dead*, *fed*, *head*, *led*, *red*, *said*, *wed; beck*, *bell*, *bet*.

In other words, it is the distinctions which are important, and it is for this reason that linguistic units have a purely relational identity. The principle is not easy to grasp, but Saussure offers a concrete analogy. We are willing to grant that in an important sense the 8:25 Geneva-to-Paris Express is the same train each day, even though the coaches, locomotive, and personnel change from one day to the next. What gives the train its identity is its place in the system of trains, as indicated by the timetable. And note that this relational identity is indeed the determining factor: the train remains the same train even if it leaves half an hour late. Indeed, it might always leave late without ceasing to be the 8:25 Geneva-to-Paris Express. What is important is that it be distinguished from, say, the 10:25 Geneva-to-Paris Express, the 8:40 Geneva-to-Dijon local, etc.

Another analogy which Saussure uses to illustrate the notion of relational identity is the comparison between language and chess. The basic units of chess are obviously king, queen, rook, knight, bishop, and pawn. The actual physical shape of the pieces and the material from which they are made is of no importance. The king may be of any size and shape, as long as there are ways of distinguishing it from other pieces. Moreover, the two rooks need not be of identical size and shape, so long as they can be distinguished from other pieces. Thus, as Saussure points out, if

27

a piece is lost from a chess set we can replace it with any other sort of object, provided always that this object will not be confused with the objects representing pieces of a different value (*Course*, 110; *Cours*, 153-4). The actual physical properties of pieces are of no importance, so long as there are differences of some kind – any kind will do – between pieces which have a different value.

Thus one can say that the units of the game of chess have no material identity: there are no physical properties necessary to a king, etc. Identity is wholly a function of differences within a system. If we now apply the analogy to language we shall be in a position to understand Saussure's paradoxical claim that in the system of a language 'there are only differences, with no positive terms' (*Course*, 120; *Cours*, 166). Normally when we think of differences we presuppose two things which differ, but Saussure's point is that signifier and signified are not things in this sense. Just as we can't say anything about what a pawn must look like, except that it will be different from knight, rook, etc., so the signifier which we represent as *bed* is not defined by any particular noises used in uttering it. Not only do the actual noises differ from one case to another, but English could be arranged so that noises now used to express the signifier *pet* were used for the signifier *bed*, and vice versa. If these changes were made the units of the language would be expressed differently, but they would still be fundamentally the same units (the same differences remain, both on the level of the signifier and on the level of the signified), and the language would still be English. Indeed, English would remain, in an important sense, the same language if the units of the signifier were never expressed in sound but only in visual symbols of some kind.

In saying this we are obviously making a distinction between units of the linguistic system on the one hand and their actual physical manifestations or realizations on the other. Before discussing this very important distinction in greater detail, it may be useful to recapitulate the line of reasoning that led us to it. We began by noting that there was no natural link between signifier and signified, and then, trying to explain the arbitrary nature of the linguistic sign, we saw that both signifier and signified were arbitrary divisions or delimitations of a continuum (a sound spectrum on the one hand and a conceptual field on the other). This led us to infer that both signifier and signified must be defined in terms of their relations with other signifiers and signifieds, and thus we reached the conclusion that if we are to define the units of a language we must distinguish between these purely relational and abstract units and their physical realizations. The actual sounds we produce in speaking are not in themselves units of the linguistic system, nor is the physical colour which we designate in calling a book 'brown' the same thing as the linguistic unit (the signified or concept) 'brown'. In both cases, and this is a point on which Saussure rightly insists, the linguistic unit is form rather than substance, defined by the relations which set it off from other units.

'LANGUE' AND 'PAROLE'

Here, in the distinction between the linguistic system and its actual manifestations, we have reached the crucial opposition between *langue* and *parole*. *La langue* is the system of a language, the language as a system of forms, whereas *parole* is actual speech, the speech acts which are made possible by the language. *La langue* is what the individual assimilates when he learns a language, a set of forms or 'hoard deposited by the practice of speech in speakers who belong to the same community, a grammatical system which, to all intents and purposes, exists in the mind of each speaker' (*Course*, 13-14; *Cours*, 30). 'It is the social product whose existence permits the individual to exercise his linguistic faculty' (Engler, 31). *Parole*, on the other hand, is the 'executive side of language' and for Saussure involves both 'the combinations by which the speaker uses the code of the linguistic system in order to express his own thoughts' and 'the psycho-physical mechanisms which permit him to externalize these combinations' (*Course*, 14; *Cours*, 31). In the act of *parole* the speaker selects and combines elements of the linguistic system and gives these forms a concrete phonic and psychological manifestation, as sounds and meanings.

If these remarks on *parole* seem somewhat confusing it is because they contain a problem, to which we shall return in Chapter Three. If the combination of linguistic elements is part of *parole*, then syntactic rules have an ambiguous status. To make *la langue* a system of forms and *parole* the combination and externalization of these forms is not quite the same as making *langue* the linguistic faculty and *parole* the exercise of that faculty, for the faculty includes knowledge of how to combine elements, rules of combination. This latter distinction, between *langue* as system and *parole* as realization, is the more fundamental, both in Saussure and in the Saussurian tradition. However, it is not essential to define here the specific characteristics of *parole* since, as Saussure makes clear, the principal and strategic function of the distinction between *langue* and *parole* is to isolate the object of linguistic investigation. *La langue*, Saussure argued, must be the linguist's primary concern. What he is trying to do in analysing a language is not to describe speech acts but to determine the units and rules of combination which make up the linguistic system. *La langue*, or the linguistic system, is a coherent, analysable object; 'it is a system of signs in which the only essential thing is the union of meanings and acoustic images' (*Course*, 15; *Cours*, 32). In studying language as a system of signs one is trying to identify its essential features: those elements which are crucial to the signifying function of language or, in other words, the elements which are functional within the system in that they create signs by distinguishing them one from another.

The distinction between *langue* and *parole* thus provides a principle of relevance for linguistics. 'In separating *langue* from *parole*', Saussure writes, 'we are separating what is social from what is individual and what is essential from what is ancillary or accidental' (*Course*, 14, *Cours*, 30). If we tried to study everything related to the phenomenon of speech we would enter a realm of confusion where relevance and irrelevance were extremely difficult to determine, but if we concentrate on *la langue*, then various aspects of language and speech fall into place within or around it. Once we put forward this notion of the linguistic system, we can then ask of every phenomenon whether it belongs to the system itself or is simply a feature of the performance or realization of linguistic units, and we thus succeed in classifying speech facts into groups where they can profitably be studied.

For example, the distinction between *langue* and *parole* leads to the creation of two distinct disciplines which study sound and its linguistic functions: phonetics, which studies sound in speech acts from a physical point of view, and phonology, which is not interested in physical events themselves but in the distinctions between the abstract units of the signifier which are functional within the linguistic system. (It is important to note here that though Saussure states unequivocally that physical sounds themselves are not part of *la langue* and thus paves the way for the distinction between phonetics and phonology as defined above, he himself uses the terms *phonetics* and *phonology* in a very different sense. I shall continue to use them in the modern sense defined here.)

The distinction between phonetics and phonology takes us back to points made earlier about the linguistic identity of the form *bed*. Phonetics would describe the actual sounds produced when one utters the form, but, as we argued above, the identity of *bed* as a unit of English does not depend on the nature of these actual sounds but on the distinctions which separate *bed* from *bet*, *bad*, *head*, etc. Phonology is the study of these functional distinctions, and 'functional' is what should be stressed here. For example, in English utterances there is a perceptible and measurable difference between the 'l-sound' which occurs before vowels (as in *lend* or *alive*) and that which occurs before consonants or at the end of words (as in *melt* or *peel*). This is a real *phonetic* difference, but it is not a difference ever used to distinguish between two signs. It is not a functional difference and therefore is not a part of the phonological system of English. On the other hand, the difference between the vowels of *feel* and *fill* is used in English to distinguish signs (compare *keel* and *kill*, *keen* and *kin*, *seat* and *sit*, *heat* and *hit*, etc.). This opposition plays a very important role in the phonological system of English in that it creates a large number of distinct signs.

The same distinction between what belongs to particular linguistic acts and what belongs to the linguistic system itself is important at other levels too, not just that of sound. We can distinguish, for example, between utterance, as a unit of *parole*, and sentence, a unit of *la langue*. Two different utterances may be manifestations of the same sentence; so once again we encounter this central notion of identity in linguistics. The actual sounds and the contextual meanings of the two utterances will be different; what makes the two utterances instances of a single linguistic unit will be the distinctions which give that unit a relational identity.

For example, if at some time Cuthbert says 'I am tired', *I* refers to Cuthbert, and understanding this reference is an important part of understanding the utterance. However, that reference is not part of the meaning of the sentence, for George also may utter the same sentence, and in his utterance *I* will refer to George. Within the linguistic system *I* does not refer to anyone. Its meaning in the system is the result of the distinctions between *I* and *you, he, she, it, we,* and *they*: a meaning which we can sum up by saying that *I* means 'the speaker' as opposed to anyone else.

Pronouns are obvious illustrations of the difference between meanings which are properties of utterances only and meanings which are properties of elements of the linguistic system. To characterize this distinction Saussure uses the terms *signification* and *valeur* ('value'). Linguistic units have a value within the system, a meaning which is the result of the oppositions which define them; but when these units are used in an utterance they have a signification, a contextual realization or manifestation of meaning. For example, if a Frenchman says 'J'ai vu un mouton' and an Englishman says 'I saw a sheep' their utterances are likely to have the same signification; they are making the same claim about a state of affairs (namely, that at a time in the past the speaker saw a sheep). However, as units of their respective linguistic systems, *mouton* and *sheep* do not have the same meaning or value, for 'sheep' is defined by an opposition with 'mutton', whereas 'mouton' is bounded by no such distinction but it is used both for the animal and for the meat. There are certain philosophical problems here which Saussure did not tackle; in particular, philosophers would want to say that what Saussure calls the signification of an utterance involves both meaning and reference. But Saussure's point is that there is one kind of meaning, a relational meaning or value, which is based on the linguistic system, and another kind of meaning or signification which involves the use of linguistic elements in actual situations of utterance.

The distinction between *langue* and *parole* has important consequences for other disciplines besides linguistics, for it is essentially a distinction between institution and event, between the underlying system which makes possible various types of behaviour and actual instances of such behaviour. Study of the system leads to the construction of models which represent forms, their relations to one another, and their possibilities of combination, whereas study of actual behaviour or events would lead to the construction of statistical models which represent the probabilities of particular combinations under various circumstances.

In our discussion of semiology in Chapter Four we shall see how the notion of *langue* has been extended to other fields. Within linguistics itself, though, study of *la langue* involves an inventory of the distinctions which create signs and of rules of combination, whereas study of *parole* would lead to an account of language use, including the relative frequencies with which particular forms or combinations of forms were used in actual speech. By separating *langue* from *parole* Saussure gave linguistics a suitable object of study and gave the linguist a much clearer sense of what he was doing: if he focused on language as a system then he knew what he was trying to reconstruct and could, within this perspective, determine what evidence was relevant and how it should be organized.

We shall consider the structure of the linguistic system in more detail at the end of this chapter, but there is one point about the concept of *la langue* which should be stressed here. Saussure's editors organized the *Course* so that it began with the distinction between *langue* and *parole*. Saussure was thus portrayed as saying that language is a confused mass of heterogeneous facts and the only way to make sense of it is to postulate the existence of something called the linguistic system and to set aside everything else. The distinction has thus seemed extremely arbitrary to many people: a postulate which had to be accepted on faith if one were to proceed. But in fact, as Saussure's notes suggest and as the sequence of argument which we have adopted here should have demonstrated, the distinction between *langue* and *parole* is a logical and necessary consequence of the arbitrary nature of the sign and the problem of identity in linguistics. In brief: if the sign is arbitrary, then, as we have seen, it is a purely relational entity, and if we wish to define and identify signs we must look to the system of relations and distinctions which create them. We must therefore distinguish between the various substances in which signs are manifested and the actual forms which constitute signs; and when we do this what we have isolated is a system of forms which underlies actual linguistic behaviour or manifestation. This system of forms is *la langue*; the attempt to study signs leads us, inexorably, to take this as the proper object of linguistic investigation. The isolation of *la langue* is not, as the published *Course* may suggest, an arbitrary point of departure but a consequence of the nature of signs themselves.

SYNCHRONIC AND DIACHRONIC PERSPECTIVES

There is another important consequence of the arbitrary nature of the sign which has also been treated by Saussure's critics as a questionable and unnecessary imposition. This is the distinction between the *synchronic* study of language (study of the linguistic system in a particular state, without reference to time) and the *diachronic* study of language (study of its evolution in time). It has been suggested that in distinguishing rigorously between these two perspectives and in granting priority to the synchronic study of language, Saussure was ignoring, or at least setting aside, the fact that a language is fundamentally historical and contingent, an entity in constant evolution. But on the contrary, it was precisely because he recognized, more profoundly than his critics, the radical historicity of language that he asserted the importance of distinguishing between facts about the linguistic system and facts about linguistic evolution, even in cases where the two kinds of facts seem extraordinarily intertwined. There is an apparent paradox here which requires elucidation.

What is the connection between the arbitrary nature of the sign and the profoundly historical nature of language? We can put it in this way: if there were some essential or natural connection between signifier and signified, then the sign would have an essential core which would be unaffected by time or which at least would resist change. This unchanging essence could be opposed to those 'accidental' features which did alter from one period to another. But in fact, as we have seen, there is no aspect of the sign which is a necessary property and which therefore lies outside time. Any aspect of sound or meaning can alter; the history of languages is full of radical evolutionary alterations of both sound and meaning. The Old English *þing* meaning 'discussion' has gradually become the modern English *thing* with a totally different meaning. The Greek θηριακὸς (theriakos) meaning 'pertaining to a wild animal'

35

became the modern English *treacle*. The Latin *calidum* ('hot') became the modern French *chaud* (pronounced ʃo, as in English *show*), in which meaning persists but none of the original phonological elements are preserved. In short, neither signifier nor signified contains any essential core which time cannot touch. Because it is arbitrary, the sign is totally subject to history, and the combination at a particular moment of a given signifier and signified is a contingent result of the historical process.

The fact that the sign is arbitrary or wholly contingent makes it subject to history but also means that signs require an ahistorical analysis. This is not as paradoxical as it might seem. Since the sign has no necessary core which must persist, it must be defined as a relational entity, in its relations to other signs. And the relevant relations are those which obtain at a particular time. A language, Saussure says, 'is a system of pure values which are determined by nothing except the momentary arrangement of its terms' (*Course*, 80; *Cours*, 116). Because the language is a wholly historical entity, always open to change, one must focus on the relations which exist in a particular synchronic state if one is to define its elements.

In asserting the priority of synchronic description, Saussure is pointing out the irrelevance of historical or diachronic facts to the analysis of *la langue*. Some examples will show why diachronic information is irrelevant. In modern English the second person pronoun *you* is used to refer both to one person and to many and can be either the subject or object in a sentence. In an earlier state of the language, however, *you* was defined by its opposition to *ye* on the one hand (*ye* a subject pronoun and *you* an object pronoun) and to *thee* and *thou* on the other (*thee* and *thou* singular forms and *you* a plural form). At a later stage *you* came to serve also as a respectful way of addressing one person, like the modern French *vous*. Now in modern English *you* is no longer defined by its opposition to *ye*, *thee* and *thou*. One can know and speak modern English perfectly without knowing that *you* was once a plural and objective form, and indeed if one knows this there is no way in which this knowledge can serve as part of one's knowledge of modern English. The description of modern English *you* would remain exactly the same if its historical evolution had been wholly different, for *you* in modern English is defined by its role in the synchronic state of the language.

Similarly, the French noun *pas* ('step') and the negative adverb *pas* ('not') derive historically from a single sign, but this is irrelevant to a description of modern French, where the two words function in totally different ways and must be treated as distinct signs. It makes no difference to modern French whether these two signs were once, as is in fact the case, a single sign, or whether they were once totally distinct signs whose different signifiers have become similar through sound changes (this has happened, for example, with the English *skate*, where sound changes have brought together the fish *skate*, from Old Norse *skata*, and ice *skate*, from *Dutch schaats*). To try to incorporate these historical facts into an account of the contemporary linguistic system would be a distortion and falsification.

Saussure's insistence on the difference between synchronic and diachronic perspectives and on the priority of synchronic description does not mean, however, that he had deceived himself into thinking that language exists as a series of totally homogeneous synchronic states: English of 1920, English of 1940, English of 1960. In a sense, the notion of a synchronic state is a methodological fiction. When we talk about the linguistic system of French at a given time we are abstracting from a reality which consists of a very large number of native speakers, whose linguistic systems may differ in various ways. Nevertheless, the linguistic system of French is a definite reality, in that all these speakers understand one another, whereas someone who speaks only English cannot understand them. Since we want to represent this fact and speak of the system which these native speakers have in common, we produce statements about the linguistic system in a particular synchronic state.

Moreover, even if the notion of a synchronic state is a methodological fiction, it is important to remember that statements about the historical evolution of language are equally fictitious. Suppose I wished to make the diachronic claim that in twentieth-century French the sound /ɑ/ has become /a/ (I follow here the convention of placing phono-logical forms between oblique lines). What does this mean? To say that /ɑ/ became /a/ suggests the transformation of an object in time, but in fact this is a historical fiction which summarizes a lot of synchronic facts: that at an earlier point in the century there were lots of speakers who distinguished between two *a*'s, as in *pâte* and *patte* or *tâche* and *tache*, whereas now there are few speakers who make the distinction, so that there is coming to be only one *a* in the language. Even this, of course, may be an oversimplification in that some speakers will hear the distinction but not use it themselves, whereas others will use it only in relatively formal circumstances.

As this example shows, a diachronic statement relates a single element from one state of a linguistic system to an element from a later state of the system. Given the relational nature of linguistic units, the fact that they are wholly defined by relations within their own state of the system, this is a questionable thing to do, foreign to the principle of synchronic linguistics. How can it be justified? How can one postulate a diachronic identity?

Saussure argues that despite their different status, dia-chronic statements are derived from synchronic statements. What allows us, he asks, to state the fact that Latin *mare* became French *mer* ('sea')? The historical linguist might argue that we know *mare* became *mer* because here, as else-where, the final *e* was dropped and *a* became *e*. But, Saussure argues, to suggest that these regular sound changes are what create the link between the two forms is to get things back-wards, because what enables us to identify this sound change is our initial notion that one form became the other. 'We are using the correspondence between *mare* and *mer* to decide that *a* became *e* and that final *e* fell' (*Course*, 182; *Cours*, 249).

In fact, what we are supposing in connecting *mare* and *mer* is this: that *mare*, *mer*, and the intermediate forms constitute an unbroken chain of synchronic identities. At each period where, retrospectively, we can say that a change occurred, there was an old form and a new form which were phonetically different but phonologically or functionally identical. They may of course have had different associations (e.g. one form might have seemed a bit old-fashioned) but they could be used interchangeably by speakers. No doubt some would stick to the old form and others prefer the new, but since the move from one to the other would not produce a difference in actual meaning, from the point of view of the linguistic system there would be a synchronic identity between the two forms. It is in this sense that diachronic identity depends on a series of synchronic identities.

As Saussure says with respect to another example, 'the diachronic identity of two words as different as *calidum* and *chaud* ("hot") means simply that one passed from the former to the latter through a series of synchronic identities' (*Course*, 182; *Cours*, 250). At one point *calidum* and *calidu* were interchangeable and synchronically identical, then later *calidu* and *caldu*, then *caldu* and *cald*, then *cald* and *tʃalt*, then *tʃalt* and *tʃaut*, then *tʃaut* and *ʃaut*, then *ʃaut* and *ʃot*, and finally *ʃot* and *ʃo* (the pronunciation of *chaud*). When we speak of the transformation of a word and postulate a diachronic identity we are in fact summarizing a parleyed series of synchronic identities. 'This is why I said', Saussure continues, 'that knowing why "Gentlemen!" retains its identity when repeated several times during a lecture is just as interesting as knowing . . . why *chaud* is identical to *calidum*. The second problem is in fact only an extension and complication of the first' (*Course*, 182; *Cours*, 250).

Thus, one cannot argue that diachronic linguistics is in

some way closer to the reality of language, while synchronic analysis is a fiction. Historical filiations are derived from synchronic identities. Not only that, they are facts of a different order. Synchronically speaking, diachronic identities are a distortion, for the earlier and later signs which they relate have no common properties. Each sign has no properties other than the specific relational properties which define it within its own synchronic system. From the point of view of systems of signs, which after all is the point of view which matters when dealing with signs, the earlier and later sign are wholly disparate.

Whence the importance of separating the synchronic and diachronic perspectives, even when the facts they are treating seem inextricably intertwined. This is a point which one must stress, because linguists who oppose Saussure's radical distinction between synchronic and diachronic approaches and wish to envisage a synthetic, panchronic perspective often point to the entanglement of synchronic and diachronic facts as if it supported their case. Saussure is all too aware of the intertwining of synchronic and diachronic facts; indeed, for him the whole difficulty is one of separating these elements when they are mixed, because only in this way can linguistic analysis attain coherence. Linguistic forms have synchronic and diachronic aspects which must be separated because they are facts of a different order, with different conditions of existence.

A panchronic synthesis is impossible, Saussure argues, because of the arbitrary nature of linguistic signs. In other sorts of systems one might hold together the synchronic and diachronic perspectives: 'insofar as a value is rooted in things themselves and in their natural relations, one can, to a certain extent, follow this value through time, bearing in mind that it depends at each moment on a system of values that coexist with it' (*Course*, 80; *Cours*, 116). Thus, the value of a piece of land at a given moment will depend on a great many other factors in the economic system, but the value is somewhat rooted in the nature of the land itself and variations will not involve simply the replacement of one arbitrary value by another. But in the case of language where the value of a sign has no natural basis or inherent limits, historical change has a different character. Elements of a language, Saussure says, are abandoned to their own historical evolution in a way that is wholly unknown in areas where forms have the smallest degree of natural connection with meaning (Engler, 169). Since no signifier is naturally more suited to a signified than any other, sound change takes place independently of the system of values: 'a diachronic fact is an event with its own rationale; the particular synchronic consequences which may follow from it are completely foreign to it' (*Course*, 84; *Cours*, 121).

Saussure's argument here is a complicated one. The claim is that diachronic facts are of a different order from synchronic facts in that historical change originates outside the linguistic system. Change originates in linguistic performance, in *parole*, not in *la langue*, and what is modified are individual elements of the system of realization. Historical changes affect the system in the end, in that the system will adjust to them, make use of the results of historical change, but it is not the linguistic system which produces them.

One thing Saussure is opposing here is the notion of teleology in linguistics: the idea that there is some end towards which linguistic changes are working and that they take place in order to achieve that end. Changes do not occur in order to produce a new state of the system. What happens is that 'certain elements are altered without regard to their solidarity within the system as a whole.'[2] These isolated changes have general consequences for the system in that its network of relations will be altered. However, 'it is not that one system has produced another but that an element of the first has been changed, and that has sufficed to bring into existence another system' (*Course*, 85; *Cours*, 121). Changes are part of an independent evolutionary process to which the system adjusts.

A diachronic fact involves the displacement of one form

by another. This displacement does not in itself have any significance; from the point of view of the linguistic system, it is non-functional. A synchronic fact is a relationship or opposition between two forms existing simultaneously: a relationship which is significant in that it carries meaning within the language. Whenever linguistic change has repercussions for the system one will have a situation where both sorts of facts are mixed together and are easy to confuse. But they are very different and must be separated. In order to grasp the difference and its importance, let us consider some English nouns with unusual plural forms: *feet, geese, teeth*. What are the synchronic and diachronic aspects of the development of these forms?

In early Anglo-Saxon the singular and plural forms of these nouns seem to have been as follows:

Stage One
	singular	plural	
foot:	fōt	fōti	(pronounced roughly *foat, foati*)
goose:	gōs	gōsi	
tooth:	tōþ	tōþi	(where þ = th)

Then plural forms were affected by a phonetic change known as '*i* mutation': when *i* followed a stressed syllable the vowel of the stressed syllable was affected and back vowels were fronted, so that *ō* became *ē*. This gave

Stage Two
	singular	plural
foot:	fōt	fēti
goose:	gōs	gēsi
tooth:	tōþ	tōþi

Then in a second phonetic change the final *i* was dropped, giving

Stage Three
singular plural

foot: fōt   fēt
goose: gōs   gēs
tooth: tōþ   tēþ

These forms, by the Great English Vowel Shift in which ō became ū and ē became ī, then became the modern forms (*Course*, 83-4; *Cours*, 120).

At stage one the plural was marked by the presence of a final *i*. This is a synchronic fact: that the opposition between presence and absence of *i* marked the opposition between singular and plural. Then a phonetic change, which had nothing to do with plurals or indeed with the grammar of the language at all, brought about a change in those forms containing a final *i*. This change had nothing to do with plurals (nothing to do with the synchronic opposition between singular and plural) in that it occurred wherever *i* followed a stressed syllable – even in verbs, for example. But as it happened, a certain number of plural forms were affected, producing a new synchronic fact in stage two. Some plural forms, as a result of an event which had nothing to do with plurals as such, had come to be marked by a double opposition: between the presence and absence of a final *i*, as before, and between the *e* of the plural and the *o* of the singular. Then, with the fall of the final *i*, which again did not concern plurals as such, there came to be a new synchronic situation. The shape of the plural forms had changed through a historical event, but since there was still a difference between the singular and plural forms (*o* as opposed to *e*) the linguistic system was able to use this difference as a meaning-bearing opposition.

'This observation', Saussure writes,

helps us to understand more fully the fortuitous nature of a linguistic state ... The state which resulted from the change was not designed to signal the meanings with which it has been endowed. A fortuitous state was given (*fōt*:*fēt*); and speakers took it over to make it carry the distinction between singular and plural. *Fōt: fēt* is no better suited to this purpose than *fōt: fōti*. In each state mind breathes life into a substance which is given (*Course*, 85; *Cours*, 121-2).

From the point of view of the linguistic system, the significant facts are the synchronic ones. Diachronic events throw up new forms which then become part of a new system, but, as Saussure says, 'in the diachronic perspective one is dealing with phenomena which are unrelated to the systems, though they condition them' (*Course*, 85; *Cours*, 122).

Saussure urges the necessity of distinguishing the synchronic and diachronic perspectives in all cases, but his discussion treats only sound changes. Of course, the examples he discusses do have morphological and grammatical consequences within the system, and such readjustments may eventually have semantic consequences, but he never deals with the problem of semantic change itself, the diachronic alterations of signifieds. He admits in passing that once one leaves the plane of sound it becomes more difficult to maintain the absolute distinction between the synchronic and the diachronic (*Course*, 141; *Cours*, 194); but the theory certainly enjoins one to do so, and one can make out a plausible though unfashionable case for the extension of the distinction to semantics.

The argument is formally very similar to that involving sound changes. Suppose one were studying the change in meaning of *kunst* in Middle High German between roughly 1200 and 1300. What would be synchronic and what diachronic here? To define change of meaning one needs two meanings and these can only be determined by considering synchronic facts: the relations between signifieds in a given state of the language which define the semantic area of 'kunst'. At an early stage it was a higher, courtly knowledge or competence, as opposed to lower, more technical skills ('list'), and a partial accomplishment as opposed to the synoptic wisdom of 'wîsheit'. In a later stage the two major oppositions which defined it were different: mundane versus spiritual ('wîsheit') and technical ('wizzen') versus non-technical. What we have are two different organizations of a semantic field. A diachronic statement would be based on this synchronic information, but if it were to explain what happened to 'kunst' it would have to refer to non-linguistic factors or causes (social changes, psychological processes, etc.) whose effects happened to have repercussions for the semantic system. For analysis of the language the relevant facts are the synchronic oppositions. The diachronic perspective treats individual filiations, which are identifiable only from the results of synchronic analysis, and draws upon what Stephen Ullmann calls 'the infinite variety and complexity of causes which govern semantic change' in order to account for the move from one state to another. But a knowledge of previous meanings and of the particular causes of change would not be relevant to an account of the semantic relations of a synchronic state (except insofar as previous meanings were still present in the system, in which case they would be considered synchronically, not diachronically).

Here, as in the cases which Saussure considers, diachronic facts are of a different order from the synchronic, bearing on individual elements rather than on the system which alone can define those elements as linguistic units. History, the historical evolution of individual elements, throws up forms which the system uses, and study of those systematic uses is the central task. Historical or causal explanation is not what is required; it bears on the elements of a language, not the language, and bears on them only as elements. Explanation in linguistics is structural: one explains forms and rules of combination by setting out the underlying system of relations, in a particular synchronic state, which create and define the elements of that synchronic system.

ANALYSIS OF 'LA LANGUE'

The two major consequences of the arbitrary nature of the sign, which we have now explored, both point to what is a single fact and may be considered as the centre of Saussure's theory of language. Language is form, not substance. A language is a system of mutually related values, and to analyse the language is to set out the system of values which constitute a state of the language. As opposed to the positive phonic and signifying elements of speech acts or *parole*, *la langue* is a system of oppositions or differences, and the task of the analyst is to discover what are these functional differences.

The basic problem, as we have followed Saussure in insisting, is that of linguistic identity. Nothing is given in linguistics. There are no positive, self-defined elements with which we can start. In order to identify two instances of the same unit we must construct a formal and relational entity by distinguishing between differences which are non-functional (and hence, for Saussure, non-linguistic) and differences which are functional. Once we have identified the relations and oppositions which delimit signifiers on the one hand and signifieds on the other, we have things which we may speak of as positive entities, linguistic signs, though we must remember that they are entities which emerge from and depend on the network of differences which constitutes the linguistic system at a given time.

But so far, in speaking of signs or linguistic units, it may sound as though we were speaking of words only, as if language consisted of nothing more than a vocabulary, organized according to phonological and semantic opposi-tions. Of course language consists also of many grammatical relations and distinctions, but Saussure insists, in a passage that is worth quoting at length, that there is no fundamental difference between a linguistic unit and a grammatical fact. Their common nature is a result of the fact that signs are entirely differential objects and that what constitutes a linguistic sign (of whatever kind) is nothing but differences between signs.

'A rather paradoxical consequence of this principle is that, in the final analysis, what is commonly referred to as a "grammatical fact" fits our definition of a linguistic unit.'

It is always expressed by an opposition between terms. Thus, in the case of the German opposition between *Nacht* ('night') and *Nächte* ('nights') it is the difference which carries grammatical meaning.

Each of the terms present in the grammatical fact (the singular without umlaut and final *e*, as opposed to the plural with umlaut and final *e*) is itself the result of the interplay of oppositions within the system. Taken by itself, neither *Nacht* or *Nächte* is anything; thus everything lies in the opposition. In other words, one could express the relationship between *Nacht* and *Nächte* by an algebraic formula a/b, where a and b are not simple terms but are themselves each the result of a set of oppositions. The linguistic system is, as it were, an algebra which contains only complex terms. Among its oppositions some are more significant than others, but 'linguistic unit' and 'grammatical fact' are only different names for designating aspects of the same general phenomenon: the play of linguistic oppositions. So true is this that we could approach the problem of linguistic units by starting with grammatical facts. Taking an opposition like *Nacht*: *Nächte* we could ask what are the units involved in this opposition? Are they these two words only, or the whole series of similar words? or a and $ä$, or all singulars and plurals, etc.?

Linguistic unit and grammatical fact would not be similar to one another if linguistic signs were made up of something besides differences. But the linguistic system being what it is, wherever one begins one will find nothing simple but always and everywhere this same complex equilibrium of reciprocally defined or conditioned terms. In other words, language is a form and not a substance. One cannot steep oneself too deeply in this truth, for all the mistakes in our terminology, all our incorrect ways of designating aspects of language, come from this involuntary assumption that linguistic phenomena must have substance (*Course*, 121-2; *Cours*, 168-9).

Consider, for example, the case of the English word *took*. What is the sign of the past tense here? It is obviously nothing positive in the word itself but a relational element. The opposition between *take* and *took* carries the distinction between present and past, just as the opposition between *foot* and *feet* carries the distinction of number. Without *feet*, *foot* would presumably be indeterminate, just as *sheep* is (cf.

'I saw the sheep in the field'). Grammatical facts illustrate the purely relational nature of the sign and confirm Saussure's radical conception of the 'fundamentally identical nature of all synchronic facts' (*Course*, 134; *Cours*, 187).

In studying a language, then, the linguist is concerned with relationships: identities and differences. And he discovers, Saussure argues, two major types of relationship. On the one hand, there are those which we have so far been discussing: oppositions which produce distinct and alternative terms (*b* as opposed to *p*; *foot* as opposed to *feet*). On the other hand, there are the relations between units which combine to form sequences. In a linguistic sequence a term's value depends not only on the contrast between it and the others which might have been chosen in its stead but also on its relations with the terms which precede and follow it in sequence. The former, which Saussure calls *associative* relations, are now generally called *paradigmatic* relations. The latter are called *syntagmatic* relations. Syntagmatic relations define combinatory possibilities: the relations between elements which might combine in a sequence. Paradigmatic relations are the oppositions between elements which can replace one another.

These relations hold at various levels of linguistic analysis. The phoneme /p/ in English is defined both by its opposition to other phonemes which could replace it in contexts such as /–et/ (cf. *bet, let, met, net, set*), and by its combinatory relations with other phonemes (it can precede or follow any vowel; within a syllable the liquids /l/ and /r/ are the only consonants which can follow it and /s/ the only one that can precede it).

At the level of morphology or word structure we also find both syntagmatic and paradigmatic relationships. A noun is partly defined by the combinations into which it can enter with prefixes and suffixes. Thus we have *friendless, friendly, friendliness, unfriendly, befriend, unbefriended, friendship, unfriendliness*, but not **disfriend, *friendier *friendation, *subfriend, *overfriend, *defriendize*, etc. The combinatory possibilities represent syntagmatic relationships, and the paradigmatic relationships are to be found in the contrast between a given morpheme and those which could replace it in a given environment. Thus, there is paradigmatic contrast between *-ly*, *-less*, and *-ship* in that they can all occur after *friend* and replacement of one by another brings a change in meaning. Similarly, *friend* has paradigmatic relations with *lecture, member, dictator, partner, professor*, etc. in that they all contrast with one another in the environment ___ *-ship*.

If we move up to the level of syntax proper we can continue to identify the same types of relationship. The syntagmatic relations which define the constituent *he frightened* permit it to be followed by certain types of constituent only: *George, the man standing on the corner, thirty-one fieldmice*, etc. but not *the stone, sincerity, purple, in*, etc. Our knowledge of syntagmatic relations enables us to define for *he frightened* a paradigmatic class of items which can follow it. These items are in paradigmatic contrast with one another, and to choose one is to produce meaning by excluding others.

Saussure claims that the entire linguistic system can be reduced to and explained in terms of a theory of syntagmatic and paradigmatic relations and that in this sense all synchronic facts are fundamentally identical. This is perhaps the clearest assertion of what may be called the structuralist view of language: not simply that a language is a system of elements which are wholly defined by their relations to one another within the system, though it is that, but that the linguistic system consists of different levels of structure; at each level one can identify elements which contrast with one another and combine with other elements to form higher-level units, and the principles of structure at each level are fundamentally the same.

We can summarize and illustrate this view by saying that since language is form and not substance its elements have only contrastive and combinatorial properties, and that at each level of structure one identifies the units or elements of a language by their capacity to differentiate units of the level immediately above them. We identify phonological distinctive features as the relational features which differentiate phonemes: /b/ is to /p/ and /d/ is to /t/ as voiced is to voiceless; thus voiced versus voiceless is a minimal distinctive feature. These phonemes in turn are identifiable because the contrasts between them have the capacity to differentiate morphemes: we know that /b/ and /p/ must be linguistic units because they contrast to distinguish *bet* from *pet*. And we must treat *bet* and *pet* as morphological units because the contrast between them is what differentiates, for example, *betting* from *petting* or *bets* from *pets*. Finally, these items, which we can informally call words, are defined by the fact that they play different roles in the higher-level units of phrases and sentences.

In thus asserting the mutual dependence of the various levels of language we are once again showing how it is that in linguistics nothing is given in advance. Not only that, we are arguing that one cannot first identify the elements or units of one level and then work out the way they combine to form units of the next level, because the elements with which one tries to start are defined by both syntagmatic and paradigmatic relations. The only way we can identify the prefix *re-* as a morphemic unit of English is by asking not just whether it contrasts with other elements but whether, when it combines with other elements to form a higher-level unit, it enters into contrasts which distinguish and define the higher-level combination. We know that *re-* contrasts paradigmatically with *un-*, *out-*, and *over-* because *redo* contrasts with *undo*, *outdo*, and *overdo*; and we know that *do* is a separable morphemic element because *redo* contrasts with *rebuild*, *reuse*, *reconnect*, etc. It is, shall we say, only contrasts between words which enable us to define the lower-level constituents of words, morphemes. One must simultaneously work out syntagmatic and paradigmatic relations. This basic structural principle, that items are defined by their contrasts with other items and their ability to combine to form higher-level items, operates at every level of language.

LANGUAGE AS SOCIAL FACT

In explaining these technical aspects of Saussure's theory of language we have not emphasized sufficiently one principle to which he gave great weight: that in analysing a language we are analysing social facts, dealing with the social use of material objects. As we have said, a language could be realized in various substances without alteration to its basic nature as a system of relations. What is important, indeed all that is relevant, are the distinctions and relations that have been endowed with meaning by a society. The question the analyst constantly asks is what are the differences which have meaning for members of the speech community. It may often be difficult to assign a precise form to those things that function as signs, but if a difference bears meaning for members of a culture, then there is a sign, however abstract, which must be analysed. For speakers of English *John loves Mary* is different in meaning from *Mary loves John*; therefore the word-order constitutes a sign, a social fact, whereas some physical differences between the way two speakers utter the sentence, *John loves Mary*, may bear no meaning and therefore be purely material facts, not social facts.

We can see, then, that the linguist studies not large collections of sound sequences but a system of social conventions. He is trying to determine the units and rules of combination which make up that system and which make possible linguistic communication between members of a society. It is one of the virtues of Saussure's theory of language to have placed social conventions and social facts at the centre of linguistic investigation by stressing the problem of the sign. What are the signs of this linguistic system? On what does their identity as signs depend? Asking these simple questions, demonstrating that nothing can be taken for granted as a unit of language, Saussure continually stressed the importance of adopting the right methodological perspective and seeing language as a system of socially determined values, not as a collection of sub-stantially defined elements. One might, to conclude this discussion, quote two relevant passages which he actually wrote:

The ultimate law of language is, dare we say, that nothing can ever reside in a single term. This is a direct consequence of the fact that linguistic signs are unrelated to what they designate, and that therefore *a* cannot designate anything without the aid of *b* and vice versa, or in other words that both have value only by the differences between them, or that neither has value, in any of its constituents, except through this same network of forever negative differences.

Since language consists of no substance but only of the isolated or combined action of physiological, psychological, and mental forces; and since nevertheless all our distinctions, our whole terminology, all our ways of speaking about it are moulded by the involuntary assumption that there is substance, one cannot avoid recognizing, before all else, that the most essential task of linguistic theory will be to disentangle the state of our basic distinctions. I cannot grant anyone the right to construct a theory while avoiding the work of definition, although this convenient procedure seems so far to have satisfied students of language.[3]

To promote dissatisfaction and stimulate thought about fundamentals, to insist on the relational nature of linguistic phenomena: these are the vectors of Saussure's theory. We can now consider the wider significance of his work: its relation to previous and subsequent thought about language and to work in other disciplines.

3 The Place of Saussure's Theories

There are three different contexts in which one can attempt to assess the significance of Saussure's thought; and although this may involve some repetition in setting out the importance of particular concepts or insights, it seems best to consider in turn Saussure's relation to his predecessors in linguistics, the relations between Saussure's theories of language and major currents of thought outside linguistics, and finally Saussure's influence on modern linguistics and the fortune of his ideas among his successors.

This broad panorama is necessary because the importance of Saussure lies not simply in his contribution to linguistics *per se* but in the fact that he made what might otherwise have seemed a recondite and specialized discipline a major intellectual presence and model for other disciplines of the 'human sciences'. In other words, the implicit claim of this chapter is that in looking at the way in which Saussure responded to the state of linguistics in his day and at the theoretical basis on which he proposed to renovate linguistics, we shall discover fundamental insights about the study of human behaviour and social objects.

LINGUISTICS BEFORE SAUSSURE

The *Course in General Linguistics* begins with a compressed version of Saussure's remarks on the history of linguistics. Setting aside the study of language prior to 1800, he distinguishes two stages of linguistic investigation: that of comparative philology or comparative grammar, which dates from Franz Bopp's work of 1816 (which compared the conjugation system of Sanskrit with that of other languages), and a second period beginning roughly in 1870 when comparative philology became more properly histor-

53

ical and when some linguists began asking pertinent questions about the nature of language and linguistic method.

Of linguistics before 1800 Saussure has very little to say, probably because he was much less concerned with general problems of intellectual history than with methods of linguistic analysis and the definition of linguistic facts. But if we are thinking about the wider significance of Saussure's own theory we cannot avoid considering how far Saussure's revolt against the linguistics of his own century involved a dialectical working out of some of the underlying principles or implications of linguistic study prior to the nineteenth century. Our account will necessarily be sketchy, selective, and abstract, but it is essential if we are to see what Saussure rediscovered or preserved of previous thought about language.

Anyone choosing to devote himself to the study of language assumes that he is undertaking something worthwhile, and though he may not necessarily have formulated a view about the purpose of studying language, the assumptions on which his work and that of his contemporaries is based will necessarily shape their discipline. An age which assumes, for example, that linguistics will give insight into the characteristics of the nation or race will produce a very different discipline from one which assumes that linguistics will cast light on the nature of human thought and of the mind itself.

This latter assumption structured and animated the study of language in the seventeenth and eighteenth centuries: by studying language one sought to understand thought itself. But the study of language takes two different forms, according to the type of question asked about thought. The first approach, which is essentially of the seventeenth century and is best represented by the Port Royal Grammar (*Grammaire générale et raisonnée*), takes language as a picture or an image of thought and therefore seeks through a study of language to discover a universal logic, the laws of reason. The main enterprise is a rational explanation of the parts of speech and grammatical categories. Thus we are told, for example, that the verb is essentially a representation of affirmation, so that the true universal verb is *to be*. Languages have, however, joined in their verbs the truly verbal function of affirmation or predication and the other non-verbal function of designating an attribute. *Peter lives* is analysed in logical grammar as *Peter is living*, where the true verb, *is*, predicates the attribute *living* of Peter.

Grammar of this kind was wholly a-temporal or synchronic. Saussure himself, asking rather insultingly 'how did those who studied language before the foundation of linguistics proceed?', noted that the seventeenth-century grammarians' point of view was 'irreproachable'. They had a well-defined object, knew what they were doing, and did not confuse synchronic and diachronic studies, though their practice was wanting in many other respects (*Course*, 82; *Cours*, 118). But it was precisely this absence of a temporal dimension which worried their eighteenth-century successors. If one wishes to understand thought, they suggested, it is not enough to work out a logical grammar; one must discuss the formation or development of ideas. For followers of Locke this was especially crucial: to understand the human mind one must know how ideas are developed from sensations, and it was precisely this problem which the eighteenth-century *savant* and linguist, Condillac, addressed in his *Essay on the Origin of Human Knowledge*.

Condillac set out to demonstrate that reflection can be derived from sensation and that the mechanism of derivation is a 'linking of ideas' brought about through the use of signs. The precise nature of his argument is not important; what is important is the direction in which it leads him. Trying to show that thought has a natural origin, that the existence of reflection and abstract notions is something which can be explained, he went beyond the claim that language is a picture of thought (the seventeenth-century position) to argue that abstract ideas are a result of the process by which signs are created. He had therefore to demonstrate that there was a natural process by which a language of conventional signs could arise from a primitive

55

and non-reflective experience. He had to concern himself with the origin of language.

Through Condillac and his followers the origin of language became a central problem of eighteenth-century thought, but it is essential to note that it was investigated as a philosophical rather than a historical problem. One worked on the origin of language in order to shed light on the nature of language and thus on the nature of thought. By explaining the origin of something one explained its nature. And thus eighteenth-century thinking about language came to focus especially on what one might call philosophic etymology: the attempt to explain signs and abstract ideas by imagining their origins in gesture, action, and sensation. Condillac suggests, for example, that prepositions were originally the names of gestures indicating direction. The reason for such hypotheses is perhaps amply indicated in Locke's suggestion, taken up at length in Turgot's article on etymology in the French *Encyclopédie*, that study of the origins of words would indicate the concepts which Nature itself has suggested to men.

The desire to study in language the mechanism of the mind led to a search for primitive roots: essential elements which, with their meaning, lie at the core of all signs which have since developed from them. A root was a rudimentary name, a basic representation, and later developments could be thought of as metaphorical extensions or accretions, if not distortions, of these basic signs. The derivations in Horne Tooke's *The Diversions of Purley* are only the most amusing examples of a mode of thought extremely common in England and France in the eighteenth century.

Here is Tooke on the root *bar*:

A *bar* in all its uses is a *defence*: that by which anything is fortified, strengthened or defended. A *barn* is a covered enclosure in which the grain, etc. is protected or defended from the weather, from depredations, etc. A *baron* is an armed, defenceful, or powerful man. A *barge* is a strong boat. A *bargain* is a confirmed, strengthened agreement . . . A *bark* is a stout vessel. The *bark* of a tree is its defence . . .

This, as I say, is an extreme example, but it illustrates several important points. First of all, the study of language is founded on a notion of representation; it is because words are taken as signs which represent fundamental categories of experience that they are of interest, and it is according to these categories that they are grouped. Unity of representation or meaning is what is used to bring these words together.

Secondly, in order to cast light on thought the analyst attempts to motivate signs: *baron* is not simply the arbitrary combination of a phonological sequence and a meaning; it is motivated by its supposed derivation from a primitive root which is in itself the natural basis of all related signs. In general, the etymological project assumes that the words of our language are not arbitrary signs but have a rational basis and are motivated by resemblance to a primitive sign.

Thirdly, time is here invoked, as so often in the eighteenth century, not in the interests of any historical project but as an explanatory fiction. This of course opened the way for a more accurate historical study of linguistic evolution which would, by destroying philosophic etymologies, strike at the heart of the philosophic project. In invoking history, albeit as a fiction, eighteenth century students of language made themselves especially vulnerable.

Finally, the relationship between language and mind is conceived atomistically. It is when they are taken individually, or in individual groups, that signs illustrate the nature of the mind and mental operations. The connection between language and mind is here being made not through the logical structures of seventeenth-century philosophical grammar but through natural concepts represented by individual roots.

Nineteenth-century linguistics would reject these four concerns or procedures. As Hans Aarsleff argues,

it is universally agreed that the decisive turn in language study occurred when the philosophical, a priori method of the eighteenth century was abandoned in favour of the historical, a posteriori method of the nineteenth. The former began with mental categories and sought their exemplification in language, as in universal grammar, and based etymologies on conjectures about the origin of language. The latter sought only facts, evidence, demonstration; it divorced the study of language from the study of mind.[1]

Rejecting the link between language and mind, the nineteenth century lost interest in the word as a sign or representation. The word became a form which was to be compared with other forms so as to establish the relations between languages, or else a form whose historical evolution was to be traced. The fictional history of the philosophic etymologies was abandoned for a properly positivistic history, and with it was abandoned the attempt to use history to motivate signs. The object of study for nineteenth-century linguistics, in short, was no longer the sign as a representation whose rational basis must be discovered, but the form whose resemblances and historical links with other forms must be demonstrated.

Though linguists generally see nineteenth-century developments as a great advance, something was obviously lost in this shift in interest; and when Saussure came to take issue with his immediate predecessors, he returned, albeit at a different level of sophistication and in a different way, to the concerns of the eighteenth century. First of all, he returned to the problem of the sign and once again conceived of language as an order of representation. He saw that unless one treats linguistic forms as signs one cannot define them; but by placing the problem of the sign in the context of his methodological enquiry, he avoided the atomism of his eighteenth-century predecessors: signs are constituted only by their relations with other signs, so the project of studying individual signs as representations must be abandoned. Moreover, Saussure re-establishes, at least implicitly, the relationship between the study of language and the study of mind, but once again at another level and in a different methodological context. What the study of language reveals about mind is not a set of primitive conceptions or natural ideas but the general structuring and differentiating operations by which things are made to signify. When Saussure argues that meaning is 'diacritical' or differential, based on differences between terms and not on intrinsic properties of terms themselves, his claim concerns not language only but the general human process in which mind creates meaning by distinguishing.

One might say, summarizing very briefly, that eighteenth-century linguistics was an example of misplaced concreteness. The link between language and thought was made in too direct, too concrete a fashion: through individual signs whose autonomy was assumed. In order to come back to the problem in a different perspective, in order to see that it was the general mechanisms of language as a semiotic system which illustrate the properties of mind, the link between language and mind had to be broken for a time and language had to be studied as an object in itself. It had to be treated, temporarily, as a system of forms with no special relation to mind. This was the role of nineteenth-century linguistics, to which we can now turn.

In discussing the development of 'comparative grammar' or comparative philology in the nineteenth century, Saussure notes that it might not have taken place, at least not so swiftly, without Europe's discovery of Sanskrit. English rule in India and the interest taken in Indian languages by English administrators brought to the attention of European linguists the surprising affinities between Sanskrit and early European languages such as Greek and Latin. These relationships between verbal roots and between grammatical forms seemed to late eighteenth-century linguists too numerous to be fortuitous and led them to postulate a common source for the three languages.

Sanskrit encouraged the comparison of languages because, as Saussure shows, it not only possessed affinities with other Indo-European languages but helped to make clear the relations between those languages themselves. Consider the declensions of nouns below:

59

Latin:	genus	generis	genere	genera	generum
Greek:	génos	géneos	génei	génea	géneōn
Sanskrit:	ǵanas	ǵanasas	ǵanasi	ǵanassu	ǵanasām

If the Latin and Greek alone are compared with one another, there does not seem to be much direct affinity; but when the Sanskrit is added it helps to suggest the nature of the relationship between them: where the Sanskrit has an *s* between two vowels, the Latin has an *r* and the Greek has no consonant. There are still, of course, considerable differences between the vowels, but the comparison of these grammatical forms – the inflectional endings of nouns – certainly suggests strong affinities.

Faced with this new and revealing data, the task of linguistics became comparison, but not the comparison of isolated forms which had so intrigued eighteenth-century linguists. The aim was to find patterns of affinity, not to discover a primitive meaning or representation which a root like *bar* might bear in all its manifestations. And so emphasis fell on inflectional systems – precisely the elements which philosophic etymologists stripped away to get at the roots or else treated as detachable elements which were themselves derived from other roots. Friedrich von Schlegel, in his work of 1808 *On the Language and Wisdom of the Indians*, conceded the existence of common roots but argued that 'the decisive point, however, which will clarify everything here, is the inner structure of the languages or comparative grammar, which will give us entirely new information about the genealogy of language in the same way as comparative anatomy has shed light upon natural history.'

It was, as I suggested above, necessary to break the connection between the study of language and the study of mind in order to approach a better understanding of language as a system. The shift of attention from roots to inflectional patterns (which had always been the most difficult items for philosophic etymologists to deal with) reflects a change in the notion of what language is: no longer is it simply a representation, a series of forms ordered by the rationality they represent and through which one moves to grasp thought and the processes of mind itself. It is a system of forms which are governed by their own law, which possess an autonomous formal pattern. The idea of comparing languages not in terms of the roots which they use to express the fundamental concepts or categories of experience but in terms of the formal patterns of grammatical elements through which words are linked and differentiated is a major step towards the notion of a language as a formal and autonomous system.

Indeed, as Schlegel suggested in the sentence quoted above, language was now conceived as an object of knowledge, something which could be dissected or anatomized like a plant or an animal. No longer was it being studied as the form of thought itself, as a representation of the mind's relation to the world.

From the nineteenth century language began to fold in upon itself, to acquire its own particular density, to deploy a history, an objectivity, and laws of its own. It became one object of knowledge among others, on the same level as living beings, wealth and value, and the history of events and men . . . To know language is no longer to come as close as possible to knowledge itself; it is merely to apply the methods ot understanding in general to a particular domain of objectivity.[2]

The method was that of comparison; the goal was the demonstration of affinities; and the fundamental methodological principle was that analogies between inflectional systems were the criterion of linguistic relationship. But comparative study had striking results. It led to the formulation of what were called 'sound laws': general rules or tables of correspondence which stated that a particular set of sounds in one language corresponded to another set of sounds in another language. The most famous of these is Grimm's law, named after Jacob Grimm who, with Bopp, Schlegel, and Rasmus Rask, was one of the foremost comparative grammarians. Grimm's law is actually a series of nine correspondences: Germanic languages have a *t* where Latin, Greek, and Sanskrit have a *d*; an *f* where they have a *p* (these two correspondences are illustrated by words for 'foot': early Germanic *fotus*, as opposed to Greek *podos*, Latin *pedes*, and Sanskrit *padas*); that Germanic has a *b* where Latin has an *f*, Greek a *ph*, and Sanskrit a *bh*; and so on, for six other correspondences.

Saussure says that these comparative grammarians never succeeded in founding a true linguistics because they did not try to determine the nature of the object they were studying and did not ask what was the significance of the relationships they discovered (*Course*, 3; *Cours*, 16). Their method was exclusively comparative rather than historical. They spoke as if there were an abstract universal pattern, a series of slots which each language had to fill with some elements, and they thus confused the synchronic and diachronic perspective. In fact, the parallels they discovered between languages indicate a historical relationship, and the diachronic task would be to reconstruct in detail the steps by which elements of an original Indo-European language became the elements of Sanskrit, Greek, Latin, etc. The synchronic task, on the other hand, would be to show how, at a particular stage in the development of a language, the fortuitously given historical elements were organized into a system peculiar to that language.

The confusion of these two tasks, says Saussure, can be seen in Grimm, who is not a proper historical linguist. He fails to distinguish diachronic changes from the synchronic function given to new elements by the linguistic system. We saw in the previous chapter (p. 43) that vowel alternation, as in *foot: feet, goose: geese, tooth: teeth*, was the result of a purely phonetic change that did not concern grammar. But Grimm sees vowel alternation as naturally significant in itself: the vowel of *foot* becomes the vowel of *feet* in order to represent the plural (Engler, 15). It is as if there were a role which had to be filled and the language had grown or developed a new part in order to fill it. Saussure thought this kind of plausible but woolly thinking most insidious.

There was, of course, a reason for this kind of thinking: a

prestigious model to which linguists were implicitly appealing. This was the model of the living organism: a self-contained entity which grew and developed according to general laws. Schlegel, in the passage quoted earlier, related comparative grammar to comparative anatomy, and the metaphor is not uncommon in linguistic writings. Comparative anatomy, presiding over the transformation by which natural history became biology, directed research towards the inner organic structure of living beings. Plants or animals could then be related to one another in terms of the different ways in which their organisms fulfilled basic functions, such as respiration, reproduction, digestion, locomotion, circulation. These relations, in turn, led to the production of historical taxonomies: evolutionary schemes in which the notion of history could be used to bring together and explain the differences between the organic system of each species, as revealed by comparison.

The common ground between linguistics and biology in the early nineteenth century is this: both were engaged in breaking away from the fictional historical continuity which animated eighteenth-century research. The only way to do proper history was to break with history in the first instance, to treat individual languages or species as autonomous entities which could be described and compared with one another as wholes. Then, given these individual organisms, it became possible to rediscover history but at a new level. Once the living being has been analysed as an organism which finds ways of fulfilling basic functions, it has been analysed in terms of the conditions which enable it to have a history. That is to say, the history of the organism or of the species becomes the story of the way in which these basic functions are fulfilled, the story of the changes which it undergoes in order to maintain its existence. The elementary functions become the basis of a historical series. Thus, a-historical work of comparative anatomy is what made Darwin's theory of evolution possible.

Similarly, in the case of language the comparative

63

method breaks with philosophic etymology in order to consider languages as comparable systems. Comparison shows the different ways in which languages fulfil similar functions (e.g. the different inflectional systems for nouns). The analogies between these differences then call for historical explanation, for the postulation of an evolutionary tree. But here linguistics seems to have taken Lamarkian rather than Darwinian evolutionary theory as a model, with the result that languages were conceived of as evolving in a purposeful way, deliberately adapting to changes. There was a confusion between the synchronic facts – the use to which changed forms were put by the grammatical system – and diachronic facts – the sound changes themselves.

But to state the comparison in this way, as if biology were a bad influence, is no doubt unfair to biology, for Darwin himself stated quite plainly the principle which Saussure saw as essential to a proper understanding of linguistics. Any purposiveness in biological evolution, Darwin saw, does not lie in changes themselves but wholly in the process of natural selection, which is, in a sense, a synchronic process. New species develop from accidental or random mutations, which themselves have no direction or purpose. But some mutants fare better than others in the ecological system of a particular moment. The failures die, the successes persist within the system; and thus a change in the species has been brought about. But mutations, like vowel changes, do not occur *in order* to bring about a better-adapted species. Change in the species is a use to which mutations are put by the system. It is a result of mutation but, like synchronic facts, is not the goal or purpose of the original event.

THE NEO-GRAMMARIANS

It was only towards 1870, Saussure wrote, that linguists began to lay the foundation for a proper study of language. There were two important developments in which Saussure himself played a significant role. First, a group of linguists now known as the 'Neo-Grammarians' and among whom were Saussure's teachers at Leipzig, demonstrated that sound laws, which previously had been treated as correspondences that held in large numbers of cases but not in others, operated *without exception*. As Hermann Osthoff and Karl Brugman wrote,

every sound change, in as much as it occurs mechanically, takes place according to laws that admit no exception. That is, the direction of the sound shift is always the same for all members of the linguistic community except where a split into dialects occurs; and all the words in which the sound subjected to the change appears in the same relationship are affected by the change without exception.

The demonstration involved the discovery that sound changes were perfectly regular if one formulated in a sufficiently precise way the phonetic environments in which changes occurred (e.g. Sanskrit *t* corresponds to early Germanic *th* if it follows an accented syllable, but, if not, it corresponds to early Germanic *d*).

This may seem a minor technical advance, but in fact the principle at stake – of change without exceptions – is crucial, for reasons which perhaps none but Saussure understood. The absolute nature of sound change is a consequence of the arbitrary nature of the sign. Since the sign is arbitrary, there is no reason for a change in sound not to apply to all instances of that sound; whereas if sounds were motivated ('naturally' expressive, like *bow-wow*) then there would be resistance, depending on the degree of motivation, and exceptions. There are no exceptions because, given the arbitrary nature of the sound and its phonetic realizations, change does not apply directly to signs themselves but to sounds, or rather, to a single sound in a particular environment. It is as if, Saussure says, a string on a piano had been tightened or loosened. When we play a melody there will be a lot of false notes, but it would be wrong to say that the first note in the third bar, the second note in the fourth bar, the first note in the sixth bar, etc. had all changed. These changes are all consequences of one change in the system of realization. 'The system of sounds is the instrument on which we articulate the words of our language; if one of these elements is modified there can be diverse consequences, but the fact in itself does not concern words which are, as it were, the melodies of our repertoire' (*Course*, 94; *Cours*, 134).

The second important development after 1870, according to Saussure, was that 'the results of comparative study were brought into historical sequence' and linguists tried to construct in detail the historical sequence which alone would explain the results of comparison (Engler, 17). Saussure himself made a major contribution to historical linguistics in his *Memoir* of 1878 on the Indo-European vowel system: a work which showed the fecundity of thinking of language as a system of purely relational items, even when working at the task of historical reconstruction.

Saussure was interested in the problem of vowel alternation in Indo-European. The question was, what vowel system must the original Indo-European language have had to account for the patterns of vowel alternation found in the known languages which derive from it. The most difficult aspect of this question was the vowel *a*. Other scholars had postulated several different *a*'s in an attempt to explain the divergent results in other languages. Saussure found their solutions unsatisfactory and argued that in addition to two *a*'s there must have been another phoneme which he could describe in formal terms: it was unrelated to *e* or *o* (which derived from the two *a*'s), it could stand alone to form a syllable, like a vowel, but it could also combine with another vowel, like a consonant. He does not try to define its substance but calls it a 'sonant coefficient' and treats it as a purely formal and relational unit in the vowel system. What makes Saussure's work so very impressive is the fact that nearly fifty years later, when cuneiform Hittite was discovered and deciphered, it was found to contain a phoneme, written *h*, which behaved as Saussure had predicted. He had discovered, by a purely formal analysis, what are now known as the laryngeals of Indo-European.[3]

Saussure had certainly proved himself an accomplished Neo-Grammarian, and on many points he admired their accomplishments. He praised them for seeing, for example, that the phenomenon known as 'false analogy' to earlier linguists was not something to be despised but an important phenomenon in linguistic evolution, especially as a counter-balance to the effect of sound change. Consider the Latin *honor*: the original form was *honos*: *honosem* (nominative and accusative). By a sound change mentioned above (p. 59) intervocalic *s* (an *s* between two vowels) became *r*, giving *honos*: *honorem*. But since there now existed other paradigms such as *orator*: *oratorem* which were apparently 'regular', a new form developed 'by analogy': *honor*.

The Neo-Grammarians were the first to recognize how important this procedure was in restructuring languages, but even they, Saussure notes, were mistaken about its true nature and confused the synchronic and the diachronic aspects (*Course*, 163; *Cours* 224). The production of a new form, Saussure argues, is a synchronic phenomenon, comparable to the creative exploitation of combinatory possibilities that occurs when from *Market* the language creates *Marketeer* on the analogy of, say, *profiteer*. For Saussure, one may recall, there is no difference in kind between morphological and syntactic combinations, so that this kind of new formation is comparable to the production of a new sentence and not an example of significant change in the language. What happens in the case we are considering is that the new form and the old form, *honor* and *honos*, continue to exist side by side as optional variants, and when eventually the old form disappears this is not a significant change but only the elimination of a variant realization. The Neo-Grammarians gave too much weight to the historical perspective and failed to recognize the systematic and grammatical (i.e. fundamentally synchronic) nature of the phenomena they were studying.

But the real fault of Saussure's contemporaries was that they failed to ask themselves fundamental questions about what they were studying: questions about the nature of language itself and its individual forms, and important methodological questions about identity in linguistics, both synchronic and diachronic. The Neo-Grammarians could not do this because they had abandoned representation as the basis of their discipline: they were not thinking about signs, and for Saussure, as we have seen, it was only by thinking about signs and their nature that one could begin to discriminate between the functional and the non-functional aspects of language and attain an appropriately relational concept of linguistic units.

The Neo-Grammarians were concerned not with signs but with forms. If one asks under what conditions linguistic forms could become an object of knowledge, become the material of a discipline, one reaches the heart of the Neo-Grammarian position. Earlier comparative linguists like Bopp had clung to meaning and representation, not as what linguistics was attempting to analyse (as had been the case for eighteenth-century philosophic etymologists) but as the condition of comparison: one looked at the words which various languages used to express a particular concept and thus used continuity of meaning as a way of bringing forms together and justifying comparison. But as soon as linguists asked what was the significance of these comparisons, they were led to try to found their discipline on a historical continuity. If similarities in form were not fortuitous, they indicated common origin, and the task became that of postulating original forms and following the historical evolution which linked original forms with later forms in an unbroken series. Whence the appositeness of biological metaphors, one of whose consequences is to exclude questions of representation. A plant does not stand for something; it is not the bearer of a meaning; it is a form which grows according to laws which must be discovered.

In fact, the Neo-Grammarians had abandoned the biological metaphors, which belonged to the mid-nineteenth century; but in rejecting these metaphors as essentially mystical they retained two of their corollaries: the neglect of questions of representation, and the assumption that their science must be based on historical continuity and must analyse historical evolution. Saussure was suspicious of notions of historical continuity and saw that study of the historical evolution of forms could easily lead to the misunderstanding and neglect of questions of linguistic function. The diachronic perspective prevents one from asking the questions which would lead to pertinent synchronic description. And so it was for him a major development when the American linguist William Dwight Whitney, working within what was still essentially the Neo-Grammarian tradition, began to raise the question of the sign. In his books *Language and the Study of Language* and *Life and Growth of Language*, Whitney argued that 'Language is, in fact, an institution', founded on social convention, 'a body of usages prevailing in a certain community', a 'treasure of words and forms', each of which 'is an arbitrary and conventional sign'. In thus stressing the institutional and conventional nature of language, Saussure says, 'he placed linguistics on its true axis' (*Course*, 76; *Cours*, 110). But Whitney did not realize the consequences and implications of this new perspective. He still asserted that linguistics must be a historical science: its task is to seek causes, to explain why we speak as we do. He vastly underestimated the task of synchronic linguistics, writing that 'a mere apprehension and exposition of the phenomena of a language – of its words, its forms, its rules, its usages: that is work for grammarians and lexicographers.' He thus demonstrated that he had no awareness of problems of definition and identity, of the relational nature of linguistic units, and generally little interest in the questions concerning foundations which obsessed Saussure.

But Whitney did prompt Saussure to further reflection, did lead him to return to the problem of the sign and see that it was only by once again making representation rather than history the basis of a discipline that one could begin to distinguish the relevant from the irrelevant, the functional from the non-functional.

Saussure returned to representation but conceived it and

employed it in a different way. Linguistics would no longer be founded on continuity of representation (a common essential meaning for a whole series of forms), as it had been for the philosophic etymologists; on the contrary, discontinuity would be the ground of representation. Meanings exist only because there are differences of meaning, and it is these differences of meaning which enable one to establish the articulation of forms. Forms can be recognized, not by their persistence in a representational or historical continuity, but by their differential function: their ability to distinguish and thus produce distinct meanings.

This fundamental perception, which is not to be found in Whitney or in Saussure's other predecessors, is of revolutionary significance. Meaning depends on difference of meaning; it is only through difference of meaning that one can identify forms and their defining functional qualities. Forms are not something given but must be established through analysis of a system of relations and differences. This notion, as we shall see in the next chapter, makes possible a way of studying human behaviour and human objects which is only today coming into its own. In reinstating representation but focusing on its discontinuities Saussure helped to lay the foundation of modern thought.

FREUD, DURKHEIM, AND METHOD

In order to understand more clearly Saussure's modernity we might abandon linguistics for a moment and place the founder of modern linguistics with his two exact contemporaries: Sigmund Freud, the founder of modern psychology, and Emile Durkheim, the founder of modern sociology. These three thinkers revolutionized the social sciences by creating for their work a new epistemological context: that is to say, they conceived of their objects of study in a different way and offered a new mode of explanation.

The initial problem for a social science is the nature and status of the facts with which it is dealing. This was a particularly acute problem in the late nineteenth century because the two principal strains of the period's philosophical heritage, German idealism and an empiricist positivism, met at one point: their tendency to think of society as a result, a secondary or derived phenomenon rather than something primary. The positivists, in a Humean tradition, distinguished between an objective physical reality of objects and events and an individual subjective perception of reality. Society could not qualify as the former and thus came to be treated as the result of feelings and actions of individuals. As Jeremy Bentham wrote, 'society is a fictitious body, the sum of the several members who compose it'. Indeed, the assumption that society is the result of individuals each acting in accordance with self-interest is the very basis of Utilitarianism. And Durkheim, criticizing his predecessors, wrote that for them 'there is nothing real in society except the individual . . . The individual is the sole tangible reality that the observer can attain.' For Hegel, on the other hand, laws, manners, customs, and the state itself are expressions of Mind as it evolves and are thus to be studied as manifestations or results, not as primary phenomena. Neither view is especially propitious to the development of social sciences.

Saussure, Durkheim, and Freud seem to have recognized that this view gets things the wrong way round. For the individual, society is a primary reality, not just the sum of individual activities nor the contingent manifestations of Mind; and if one wishes to study human behaviour one must grant that there is a social reality. Man lives not simply among objects and actions, but among objects and actions which have meaning, and these meanings cannot be treated as a sum of subjective perceptions. They are the very furniture of the world. The social significance of actions, the meanings of utterances, feelings of love, anger, guilt, etc. cannot be lightly dismissed. They are social facts. As Durkheim repeatedly stated, and his two contemporaries would have agreed, his discipline is based on the 'objective reality of social facts'.

In short, sociology, linguistics, and psychoanalytic psychology are possible only when one takes the meanings which are attached to and which differentiate objects and actions in society as a primary reality, as facts to be explained. And since meanings are a social product explanation must be carried out in social terms. It is as if Saussure, Freud, and Durkheim had asked, 'what makes individual experience possible? what enables men to operate with meaningful objects and actions? what enables them to communicate and act meaningfully?' And the answer they postulated was social institutions which, though formed by human activities, are the conditions of experience. To understand individual experience one must study the social norms which make it possible.

It is not difficult to see why this should be so. When two people meet they may act politely or impolitely, and the politeness or impoliteness of their behaviour is a social and cultural fact. But an objective description of the physical actions they performed would not be a description of a social phenomenon because it would leave out of account the social conventions which make the actions what they are. Their behaviour is meaningful only with respect to a set of social conventions: it is these conventions which make it possible to be polite or impolite; they create behaviour which must therefore be described in their terms. Similarly, making a noise is not in itself a social phenomenon, but uttering a sentence is. The social phenomenon is made possible by a system of interpersonal conventions: a language.

Saussure, Freud, and Durkheim thus reverse the perspective which makes society the result of individual behaviour and insist that behaviour is made possible by collective social systems which individuals have assimilated, consciously or unconsciously. It was Freud, Lionel Trilling says, who 'made it apparent to us how entirely implicated in culture we all are . . . how the culture suffuses the remotest parts of the individual mind', making possible a whole series of feelings and actions and even the individual's sense of identity. Individual actions and symptoms can be inter-preted psychoanalytically because they are the result of common psychic processes, unconscious defences occasioned by social taboos and leading to particular types of repres-sion and displacement. Linguistic communication is possible because we have assimilated a system of collective norms which organize the world and give meaning to verbal acts. Or again, as Durkheim argued, the reality crucial to the individual is not the physical environment but the social milieu, a system of rules and norms, of collective representations, which makes possible social behaviour.

This perspective therefore involves a special type of explanation: to explain an action is to relate it to the under-lying system of norms which makes it possible. The action is explained as a manifestation of an underlying system of representations. Whether this is still to be regarded as causal explanation varies from one case to another. In his work on suicide, perhaps his most famous sociological investigation, Durkheim claimed to offer a causal explana-tion; but he was identifying the causes of high suicide rates in a society, not explaining why particular individuals commit suicide at a given moment. Their suicides are mani-festations of the weakening in social bonds which results from a particular configuration of social norms. Freud's psychological analyses are usually presented as causal explanations, but they do not have predictive force (he is not claiming that a given sequence of events will necessarily produce certain actions or symptoms) and are perhaps best regarded as an attempt to relate actions to an underlying psychic economy. Linguistics, on the other hand, does not pretend to causal analysis: it does not try to explain why an individual uttered a particular sequence at a given moment but shows why the sequence has the form and meaning it does by relating it to the system of the language.

In each case, then, despite pretentions to causal analysis, one might say that what is being offered is a structural rather than a causal explanation: one attempts to show why a particular action has significance by relating it to the system of underlying functions, norms, and categories which makes it possible.

What is especially significant here is the move away from historical explanation. To explain social phenomena is not to discover temporal antecedents and to link them in a causal chain but to specify the place and function of the phenomena in a system. There is a move from the diachronic to the synchronic perspective, which one might speak of as an internalizing of causation: instead of conceiving of causation on a historical model, where temporal development makes something what it is, the historical results are detemporalized and treated simply as a state, a condition.

This is a complex but fundamental displacement which we have already seen at work in Saussure's insistence that actual historical change which produces forms such as *foot* and *feet* is not an important explanatory factor in our analysis of English. What is important is the state, in which the plural is marked by alternation between the two vowels. The presence of this opposition in the system is a result of a historical process, but it is the system's use of that opposition which has explanatory value. However, the most striking example of this displacement, this internalization of causality, is to be found in Freud's work on the Oedipus complex, where we have the spectacle of a mind still attracted to historical and causal explanation while at least partially aware that this is not what his new mode of analysis requires.

In *Totem and Taboo*, discussing the prohibition of incest and other social taboos, Freud postulates a historical event in primitive times: a jealous and tyrannical father, who wished to keep all the women for himself and drove away his sons as they reached maturity, was killed and devoured by the sons who had banded together. In devouring him, they sought to take on his power and his role. This 'memorable and criminal deed' was the beginning of 'social organization, of moral restrictions, and of religion', because guilt and remorse created taboos. Freud recognizes that in making this deed the historical cause of social norms and psychic complexes which still exist, he is postulating the continuity of a collective psyche, which he calls the unconscious. How otherwise could a single act continue to exercise such profound effects on humanity? Part of the explanation, Freud says, is that in our psychical economy feelings of guilt may arise from wishes as well as from actual deeds, and 'this creative sense of guilt' helps to keep the consequences of the deed alive. In fact, he admits, it is possible that the original deed never actually took place; remorse may have been provoked by the sons' fantasy of killing the father. This is a plausible hypothesis, he says, and 'no damage would thus be done to the causal chain stretching from the beginning to the present day'. In fact, the question of whether the deed really took place or not 'does not in our judgment affect the heart of the matter'. But primitive men were uninhibited. For them 'thought passes directly into action. And that is why, without laying claim to any finality of judgment, I think that in the case before us it may be safely assumed that " in the beginning was the Deed".'

Freud here appears very much in the guise of an eighteenth-century thinker, using fictions of origin to discuss the nature of a thing. What is most important, however, is his recognition that if the original deed is to serve as a true historical cause, one must postulate an underlying psychic system which, in turn, makes the deed itself unnecessary. The guilt resulting from subconscious wishes in the familial situation is itself sufficient explanation of taboos. In fact, we see Freud first recognizing that the reality of his postulated historical cause is unimportant and then turning back on himself and deducing the historical event from the psychic system: everyone has these subconscious desires, and not simply as a result of the original deed, which may not have taken place; but primitive men were uninhibited and therefore must have acted thus. The historical event is asserted, as if it were a cause, but now in fact it is inferred from the subconscious system. This is a splendid example of the tension between historical explanation and the notion of explanation in terms of a system, and it is especially instructive as a lesson in modernity because the system wins against Freud's express wishes.

Saussure, Durkheim, and Freud seem responsible for this decisive step in the development of the sciences of man. By internalizing origins, removing them from a temporal history, one creates a new space of explanation which has come to be called the unconscious. It is not so much that the unconscious replaces the historical series; rather it becomes the space where any antecedents which have an explanatory function are located. Structural explanation relates actions to a system of norms – the rules of a language, the collective representations of a society, the mechanisms of a psychical economy – and the concept of the unconscious is a way of explaining how these systems have explanatory force. It is a way of explaining how they can be simultaneously unknown yet effectively present. If a description of a linguistic system counts as an analysis of a language it is because the system is something not immediately given to consciousness yet deemed to be always present, always at work in the behaviour it structures and makes possible.

Though the concept of the unconscious as such arises in the work of Freud, it is essential to the type of explanation which a whole range of modern disciplines seeks to offer and would certainly have been developed even without Freud's aid. In fact, one could argue that it is in linguistics that the concept emerges in its clearest and most irrefutable form. The unconscious is the concept which enables one to explain an indubitable fact: that I know a language (in the sense that I can produce and understand new utterances, tell whether a sequence is in fact a sentence of my language, etc.) yet I do not know what I know. I know a language, yet I need a linguist to explain to me precisely what it is that I know. The concept of the unconscious connects and makes sense of these two facts and opens a space of explora-tion. Linguistics, like psychology and a sociology of col-lective representations, will explain my actions by setting out in detail the implicit knowledge which I myself have not brought to consciousness.

Another way of describing this fundamental step – a way whose importance will become clearer in the final chapter – would be to say that it consists of placing the 'subject' or the 'I' at the centre of one's analytical domain and then deconstructing it. 'Subject' in this context means the subject of experience, the 'I' or self which thinks, perceives, speaks, etc. Comparative and historical linguistics could be carried out without explicit reference to the subject; one could note the differences between attested forms and follow the evolution of a given form without calling upon or making use of the notion of the subject who speaks, the subject who knows a language. But Saussure puts the subject right at the centre of his analytical project. The notion of the subject becomes central to the analysis of language.

How do we identify linguistic units? Always with reference to the subject. We know that /b/ and /p/ are different phonemes because for the subject *bet* and *pet* are different signs. The opposition between /b/ and /p/ differentiates signs for the speaking subject.

What makes two utterances identical? The fact that, despite measurable physical differences, they are identical for the speaking subject. 'In order to tell to what extent a thing is a reality', at least from the point of view of synchronic analysis of *la langue*, 'it is necessary and sufficient to ask to what extent it exists in the minds of speakers' (*Course*, 90; *Cours*, 128).

In all cases where we are dealing with what Saussure calls values, that is to say with the social significance of objects and actions, the subject takes on a crucial role, in that the facts one is seeking to explain come from his intuitions and judgments. However, once the subject is in place, once he is firmly established at the centre of the analytical domain, the whole enterprise of the human sciences becomes one of deconstructing the subject, of explaining meanings in terms of systems of convention which escape the sub-

77

ject's conscious grasp. The speaker of a language is not consciously aware of its phonological and grammatical systems, in whose terms his judgments and perceptions will be explained. Nor is the subject necessarily aware of its own psychic economy or of the elaborate system of social norms which governs behaviour.

The subject is broken down into its constituents which turn out to be interpersonal systems of convention. It is 'dissolved' as its functions are attributed to a variety of systems which operate through it. As Michel Foucault writes, 'the researches of psychoanalysis, of linguistics, of anthropology have "decentred" the subject in relation to the laws of its desire, the forms of its language, the rules of its actions, or the play of its mythical and imaginative discourse.' The distinction between the subject and the world is a variable one that depends on the configurations of knowledge at a given time, and the disciplines inaugurated by Saussure, Durkheim, and Freud have chipped away at what previously belonged to the subject until it has lost its place as centre or source of meaning. As it is deconstructed, broken down into component systems which are all trans-subjective, the self or subject comes to appear more and more as a construct: the result of systems of conventions. When man speaks he artfully 'complies with language'; language speaks through him, as does desire and society. Even the idea of personal identity emerges through the discourse cf a culture. The 'I' is not something given; it comes to exist, in a mirror stage which starts in infancy, as that which is seen and addressed by others.

The problem of the subject is one to which we shall return briefly in the final chapter when we come to consider some of the implications of semiology and the ways in which people working in other disciplines have actually been influenced by Saussure and his methodological programme. So far, of course, we have not been concerned with problems of influence: there is no evidence that Durkheim, Saussure, and Freud knew anything of each other's work, and though Durkheim's influence on Saussure has often been suggested, much more important than any possible surface borrowings are the affinities between the fundamental projects of these three thinkers and in particular the epistemological configurations of the disciplines they founded.

The preceding pages will have suggested that Saussure is especially interesting and suggestive as an intellectual strategist, as a forceful thinker concerned with fundamentals of method and definition. Yet it is as the father of modern linguistics that he is principally known and we must now look at some of the things he fathered in order to see what advances his work helped produce and where his linguistic theories have proven inadequate.

INFLUENCE

Saussure's influence on modern linguistics has been of essentially two kinds. First, he provided a general orientation, a sense of the tasks of linguistics, which has been profoundly influential and indeed has seldom been questioned, so much has it come to be taken for granted as the very nature of the subject itself. For Saussure the linguist's task was to analyse a language as a system of units and relations; to do linguistics was to attempt to define the units of a language, the relations between them, and their rules of combination. This sense of the task of linguistics is not found in Saussure's predecessors, though some of them may make occasional bows in this direction. But since Saussure this has become, very nearly, the definition of linguistic investigation. Not only has descriptive and theoretical linguistics grown in order to take up the central place which Saussure assigns it, but those working in historical linguistics or socio-linguistics are compelled to use adjectives like 'historical' to show how their work departs from the central activity of the discipline. Someone who wished to take issue with Saussure's view of the task of linguistics would do so not by attacking Saussure but by challenging the idea of linguistics itself.

It is in this sense that Saussure can be called the father of modern linguistics. His most important and original contribution is a silent influence which has passed into the nature of the discipline itself. Indeed, an account of structural linguistics, as inaugurated by Saussure, can include the major schools of modern linguistics. Thus, Giulio Lepschy's *A Survey of Structural Linguistics* covers the Prague School (Roman Jakobson, Nikolai Trubetzkoy, and others), the Copenhagen School (Louis Hjelmslev and other 'Glossematicians'), the 'Functionalists' (Jakobson, Emile Benveniste, André Martinet, and some contemporary British linguists), American Structuralism (Leonard Bloomfield and his followers) and even Noam Chomsky and other transformational grammarians. It is only this last group who, as we shall see, have altered in a fundamental way the concept of linguistics as bequeathed by Saussure.

There is, however, another kind of influence worth studying, the influence of specific concepts which are not strictly original to Saussure but which he helped to promote: the distinction between *langue* and *parole*, the separation of the synchronic and the diachronic perspectives, and the conception of language as a system of syntagmatic and paradigmatic relations operating at various hierarchical levels. Many of the developments of modern linguistics can be described as investigations of the precise nature and import of these concepts. Considering them in turn, we can see that even when Saussure's original formulations have been found wanting he posed the questions which have animated modern linguistics.

A. *Langue* and *Parole*

In 1933 the British linguist Sir Alan Gardiner wrote that 'to Ferdinand de Saussure belongs the merit of having drawn attention to the distinction between "speech" and "language", a distinction so far-reaching in its consequences that in my opinion it can hardly fail, sooner or later, to become the indispensable basis of all scientific treatment of grammar.' So indispensable has it been that many linguistic disagreements can be cast in the form of disputes about the precise nature of the distinction: what belongs to *langue* and what to *parole*?

Saussure himself invokes various criteria in making the distinction: in separating *langue* from *parole* one separates the essential from the contingent, the social from the purely individual, and the psychological from the material. But these criteria do not divide language in the same way and they thus leave much room for dispute. By the first *la langue* is a wholly abstract and formal system; everything relating to sound is relegated to *parole* since English would still be essentially the same language if its units were expressed in some other way. But clearly, by the second criterion we should have to revise this view: the fact that /b/ is a voiced bilabial stop and /p/ a voiceless bilabial stop is a fact about the linguistic system in that the individual speaker cannot choose to realize the phonemes differently if he is to continue speaking English. And by the third criterion one would have to admit other acoustic features to *la langue*, since differences between accen and pronunciations have a psychological reality for speakers of a language.

Saussure's distinction has been fruitful through its very openness. In fact, the varied results achieved by applying each of these criteria reflect different ways in which language can be systematic. We can state these differences in the terms suggested by Louis Hjelmslev: *langue* and *parole* can be replaced by *schema, norm, usage,* and *parole. Parole* is simply the individual speech act and not itself part of the system. Usage is a statistical regularity: one can chart the frequency of different pronunciations or of other uses of linguistic elements. A speaker of a language has a certain freedom in his choice of usage. The norm, however, is not a matter of individual choice. It is not described statistically but represented by a series of rules: e.g. the phoneme /p/ is realized in English as a voiceless bilabial stop. Finally, the schema is the most abstract conception of structure. Here there is no reference to phonic substance.

Elements are defined in abstract relational terms: /p/ is to /b/ as /t/ is to /d/, and it is irrelevant what actual features are used to manifest these differences.

Given this four-way distinction, one could, in fact, locate the division between *langue* and *parole* at any of three points: *la langue* could consist of schema only, or of schema and norm, or of schema, norm, and usage. And disputes about the nature of *la langue* have usually been of this character. Linguists of the Prague School, for example, treated *la langue* as a combination of schema and norm. Distinguishing between phonetics and phonology, they argued that phonology should investigate which phonic differences are linked with differences in meaning but that the phonological distinctive features thus isolated should be described in articulatory terms. In Roman Jakobson's influential account of distinctive features, oppositions such as *voiced* versus *voiceless* are not abstract features only but norms of physical or phonetic realization.

Other linguists, such as Daniel Jones and his British followers, have preferred to define the phoneme as a 'family' of sounds, thus including usage within *la langue*: for them, to describe the phonological system of a language is to describe linguistic usage as well as functional norms and abstract schemas. On the other hand, Hjelmslev and the exponents of his Glossematics treat *la langue* purely as an abstract schema. For them phonetic properties are not at all involved in the way in which phonemes should be described. These disputes continue, but one might say that at least in the realm of phonology, the essential questions are posed by Saussure's distinction between *langue* and *parole*.

At the syntactic level Saussure's views on what belongs to *langue* and what to *parole* are more obscure, indecisive, and questionable. He thinks of sentences as the products of individual choice and therefore treats them as instances of *parole* rather than entities of *la langue*. One is tempted to say that he failed to distinguish between sentences themselves as grammatical forms and the utterances by which sentences are realized in speech, but the problem goes deeper than this. Set idiomatic phrases, he allows, are part of the linguistic system, and even 'sentences and groups of words built on regular patterns', but he seems unwilling to consider how far the notion of 'regular pattern' can be extended, and he concludes that on the plane of syntagmatic combinations 'there is no clear-cut boundary between facts of *la langue*, which are examples of collective usage, and facts of *parole*, which depend on the free choice of the individual' (*Course*, 125; *Cours*, 173).

Because of his failure to include sentences within the linguistic system, Saussure's conception of syntax seems exceptionally weak. Language is more than a system of inter-related units; the relations which compose it are also a system of rules, and it is this aspect that Noam Chomsky stresses in replacing Saussure's *langue* and *parole* with his own concepts of *competence* and *performance*. Competence is the underlying system of rules which a speaker has mastered, and to describe competence is to analyse a language into its elements and their rules of combination. 'Clearly', Chomsky writes, 'the description of intrinsic competence provided by the grammar is not to be confused with an account of actual performance, as de Saussure emphasized with such lucidity.' But Saussure, he continues,

regards *langue* as basically a store of signs with their grammatical properties, that is, a store of word-like elements, fixed phrases, and, perhaps, certain limited phrase types. He was thus quite unable to come to grips with the recursive processes underlying sentence formation, and he appears to regard sentence formation as a matter of *parole* rather than *langue*, of free and voluntary creation rather than systematic rule. There is no place in his scheme for 'rule-governed creativity' of the kind involved in the ordinary everyday use of language.[4]

However, it is worth noting that it is precisely because he recognized the creativity of ordinary language use that Saussure was unwilling to include sentence formation in *la langue*. He did not know how to reconcile the fact that we can produce totally new sentences with the fact that a language contains phrase types. What he lacked was a notion of rule-governed creativity: individual creativity made possible by a system of rules. He did not realize that it is possible to construct a finite set of rules which will generate structural descriptions for an infinite number of sentences. This can be done, as Chomsky says, by recursive rules: rules which can be applied over and over again, such as the rule which enables one to attach a relative clause to a noun phrase (e.g. This is the dog that chased the cat that worried the rat that ate the cheese, etc.).

Someone who knows a language can recognize whether a sentence he has never before encountered is formed in accordance with the rules of that language and can himself produce new sentences which accord with the grammar. This fact is sufficient proof that the sentence must be considered as a unit of the linguistic system. It was left to Chomsky to show how the system could account for sentence formation while at the same time accounting for the creativity of individual speakers. Saussure's inability to do this is quite understandable, and he does seem at least to have understood the nature of the problem. But his neglect of the sentence as a linguistic unit is, nevertheless, an important failure, and it is here in the area of syntax, more than anywhere else, that Saussure's approach to language has been superseded.[5]

B. Synchronic and Diachronic

Of all Saussure's distinctions this is the one which has been least clearly understood and least perceptively investigated by his successors. Though the priority of synchronic description has been accepted, there has been little attempt to clarify the basic theoretical problem Saussure posed, about what precisely belongs to the synchronic and what to the diachronic perspective in discussions of linguistic change. Many linguists have asserted that one must overcome or transcend the distinction and achieve a global, synthetic view of language, but they have not come to terms with the reasons Saussure offers for thinking this impossible; and Charles Hockett was doubtless correct to observe in his 1968 survey of linguistics, *The State of the Art*, that the problem of the relationship between synchronic and diachronic studies 'had been not so much settled as swept under the rug'.

There are two sorts of claim made by those who seek to overcome the distinction between the synchronic and the diachronic. The first is that at any moment the synchronic system contains diachronic elements: archaisms, neologisms, distinctions which are in the process of disappearing, etc. This objection is irrelevant to Saussure's point. He explicitly states that 'at every moment a language implies an established system and an evolution; at every moment it is a present institution and a product of the past.' Synchronic and diachronic are not two types of element but two approaches to language. Items which are experienced as archaic at a given moment will be so identified in a synchronic analysis, but this has nothing to do with historical investigation (it would make no difference to the synchronic description, for example, if forms which speakers felt to be archaic were really new borrowings from another language).

The other kind of objection is more apposite and interesting. Linguists of the Prague School insisted that linguistic change was not a blind force but fundamentally systematic: that it was a function of the system. And recently those working on phonology in the context of transformational grammar have taken an anti-Saussurian position. Whereas Saussure maintains that sound change takes place outside the linguistic system, with external factors affecting *parole*, other linguists now argue that sound change arises within the linguistic system itself, can operate on grammatically-defined elements, and is best described as a change in rules, not as the evolution of realization elements. For example, at one point the *k* in forms like *knowledge* was pronounced. The sound change which affected *kn* seems to have depended on grammatical structure, so that the *k* remains in *acknowledge* but not in *a knowledge*.

The evidence is not conclusive, for there are other, albeit

ad hoc ways of explaining such changes. Nor is it clear whether Saussure's opposition to teleological notions of change – that change occurs because the system 'seeks' a different state – needs to be abandoned. Certainly many changes are not explicable in teleological terms: one cannot argue that the 'inadequacy' of *fōt/fōti* led the system to seek *foot/feet* as a way of marking plurals. And it may be that the evidence cited as counter-example often results from a failure to distinguish the synchronic facts of language change from the diachronic. In general, the relation between the synchronic and the diachronic is a problem which has not been sufficiently explored, and here Saussure's position, as we tried to explain it in Chapter Two, is as clear and apposite a formulation of the central difficulties as has yet been produced.

C. Relations in the Linguistic System

Saussure asserted unequivocally, as we have seen, that language is a system of differences in which all elements are defined solely by their relations with one another. He reached this conclusion, it will be recalled, by reflecting on the nature of identity in linguistics and on the properties of the linguistic sign. From a theoretical point of view this conclusion seems irreproachable, and it has exercised considerable influence. But when one actually analyses a language it becomes extremely difficult to avoid speaking as if there were positive terms. It is difficult to analyse a language purely as a system of relations. Whether this difficulty has significant theoretical implications is not clear, but it is true to say that linguists have been more successful in investigating particular types of relations or restricted sets of relations than in treating an entire language as a purely relational system.

For example, the prominence Saussure gives to binary oppositions has borne fruit. Most work in phonology has been based on a reduction of the sound continuum to discrete elements which can be defined as the point of intersection of several distinctive features. Each distinctive feature, as Jakobson says, involves the choice between 'two terms of an opposition which displays a specific differential property' (e.g. *voiced* as opposed to *voiceless*). Indeed, Jakobson and others argue that the use of binary oppositions to describe structure is not simply a methodological device but a reflection of the nature of language itself. Binary oppositions are the most natural and economical code; they are the first operations a child learns as he begins to accede to language; and more generally they are the common denominator of all thought. Once again we see Saussure and the Saussurian tradition re-establishing links between language and thought, but at the level of fundamental structuring operations.

Syntagmatic and paradigmatic relations have also been the focus of attention for many linguists, and one could argue that the differences between the various theories of grammatical description which have grown up since Saussure's day are essentially disagreements about the nature of syntagmatic relations and ways of determining them. These disagreements are not of a sort which could be succinctly summarized here. Suffice it to say that the concept of a hierarchy of linguistic levels, in which constituents of one level (such as phonemes) combine to form constituents of the next level (such as morphemes), and in which the combinatory potential of elements serves to define them, is common to a range of descriptive theories, which differ in their judgments about the weight to be given to various factors in determining relations. Should one, for example, take similar utterances and, treating them as sequences of forms, divide them at points where they differ from one another and then study the combinations which the elements so isolated enter in other sequences? Or should one begin with a theory of the various functions which linguistic elements can perform and then identify the elements which combine to perform these functions?

It is only with the advent of Chomsky's transformational-generative grammar that the importance of syntagmatic and paradigmatic elements, as defined by Saussure, has been reduced. And even there the problem has only been displaced: various kinds of paradigmatic classes do emerge, as the classes on which rules operate, as the classes which are necessary if the rules are to operate properly. And the rules themselves are representations of what Saussure would have seen as syntagmatic relations, had he extended his account of relations to make adequate provision for syntactic processes.

Moreover, the recent work of transformational grammarians returns, though at a different level, to the view expressed by Saussure that when one thinks rigorously about combinatory processes, taking nothing for granted, one discovers that there is no essential difference between morphological combinations and other syntactic combinations. For Saussure this is an inference only; his remarks on syntax are so sketchy that they offer no support for his claim. But now, just as the discovery of Hittite confirmed Saussure's hypothesis about Indo-European vowels, so transformational grammar may demonstrate the correctness of another postulate or insight.

There is one respect, however, in which the father of modern linguistics would have been disappointed in his children. Saussure maintained that linguistics was a branch of semiology, the general science of signs and systems of signs. Linguistics belongs not to the natural sciences nor to the historical sciences but to semiology. 'For me the problem of language is above all semiological ... If we wish to discover the true nature of the linguistic system, we must first study what it has in common with other systems of the same type' (*Course*, 17; *Cours*, 34-5). This advice, this programme, has been ignored by linguists. While other Saussurian concepts have been assimilated, Saussure's ruling concept, the notion of the sign and of language as a system of signs, has been largely neglected. Linguists have paid lip service to the concept but have not allowed it to govern their analysis of a language. It can be argued that if the sign were granted the role it has in Saussure this would lead to an important reorientation of linguistics, but until the attempt is made one cannot say what its consequences would be.[6] What one can say is that linguists' failure to make the sign an object of attention has led to an anomalous situation: semiology has been embraced by people working in many different fields, but linguistics itself, which Saussure placed at the centre of semiology and to which he thought semiology would make a major contribution, remains aloof. Linguistics has developed in Saussurian ways, but to understand the context within which Saussure placed linguistics we must abandon the study of language as such and look at attempts to study other social and cultural phenomena as 'languages', as systems of signs.

4 Semiology: The Saussurian Legacy

Very few paragraphs of the *Course in General Linguistics* are devoted to semiology, and this is no doubt one of the reasons why linguists generally neglected to follow Saussure's lead in developing a general science of signs which would situate and orient linguistics. But for Saussure the semiological perspective was central to any serious study of language. 'Is it not obvious', he wrote, 'that language is above all a system of signs and that therefore we must have recourse to the science of signs' if we are to define it properly? (Engler, 47).

Language is a system of signs that express ideas and is thus comparable to the system of writing, to the alphabet of deaf-mutes, to symbolic rituals, to·forms of etiquette, to military signals, etc. It is but the most important of these systems.

We can therefore imagine *a science which would study the life of signs within society* . . . We call it *semiology*, from the Greek *semeion* ('sign'). It would teach us what signs consist of, what laws govern them. Since it does not yet exist we cannot say what it will be; but it has a right to existence; its place is assured in advance. Linguistics is only a part of this general science; and the laws which semiology discovers will be applicable to linguistics, which will thus find itself attached to a well-defined domain of human phenomena (*Course*, 16; *Cours*, 33).

Since human beings make noises, use gestures, employ combinations of objects or actions in order to convey meaning, there is a place for a discipline which would analyse this kind of activity and make explicit the systems of convention on which it rests. And, Saussure argues, if linguistics is conceived as a part of semiology there will be important consequences:

90

. . . aspects of language which may at first seem extremely important (such as the use of vocal mechanisms) will become secondary considerations if they serve only to distinguish language from other semiological systems. This procedure will not only clarify the problems of linguistics; rituals, customs, etc., will, we believe, appear in a new light if they are studied as signs, and one will come to see that they should be included in the domain of semiology and explained by its laws (*Course*, 17; *Cours*, 35).

Semiology is thus based on the assumption that insofar as human actions or productions convey meaning, insofar as they function as signs, there must be an underlying system of conventions and distinctions which makes this meaning possible. Where there are signs there is system. This is what various signifying activities have in common, and if one is to determine their essential nature one must treat them not in isolation but as examples of semiological systems. In this way, aspects which are often hidden or neglected will become apparent, especially when non-linguistic signifying practices are considered as 'languages'.

But why should linguistics, the study of one particular though very important signifying system, be thought to provide the model for studying other systems? Why should linguistics be as Saussure called it, 'le patron général' of semiology? The answer takes us back to a familiar starting point, the arbitrary nature of the sign.

Linguistics may serve as a model for semiology, Saussure argued, because in the case of language the arbitrary and conventional nature of the sign is especially clear. Non-linguistic signs may often seem natural to those who use them, and it may require some effort to see that the politeness or impoliteness of an action is not a necessary and intrinsic property of that action but a conventional meaning. But if linguistics is taken as a model it will compel the analyst to attend to the conventional basis of the non-linguistic signs he is studying.

This is not to say that all signs are wholly arbitrary. There are some intrinsic constraints on the meanings action can bear and, reciprocally, on the class of actions appropriate to express a particular meaning. It is difficult to imagine a culture where a punch on the mouth might be a friendly greeting. But within such constraints there is a whole range of actions which would serve perfectly well as friendly greetings. Within this realm of available possibilities one can speak of signs as conventional and arbitrary. In fact, Saussure writes,

every means of expression used in a society is based, in principle, on a collective norm – in other words, on convention. Signs of politeness, for instance, often have a certain natural expressivity (one thinks of the way a Chinese prostrates himself nine times before the Emperor by way of salutation), but they are none-theless determined by a rule; and it is this rule which leads one to use them, not their intrinsic value. We can therefore say that wholly arbitrary signs are those which come closest to the semio-logical ideal. This is why language, the most complex and wide-spread of systems of expression, is also the most characteristic. And for this reason linguistics can serve as a model for semiology as a whole, though language is only one of its systems (*Course*, 68; *Cours*, 100-101).

By taking linguistics as a model one may avoid the familiar mistake of assuming that signs which appear natural to those who use them have an intrinsic meaning and involve no conventions.

Why is this important? Why should one wish to stress the conventional nature of non-linguistic signs? The answer is quite simple. If signs were natural, then there would really be nothing to analyse. One would say that opening a door for a woman simply *is* polite, and that's all there is to it. But if one starts with the assumption that signs are likely to be conventional, then one will seriously seek out the conventions on which they are based and will discover the underlying system which makes these signs what they are. Just as, in linguistics, the arbitrary nature of the sign leads one to think about the system of functional differences which create signs, so in other cases one will focus on significant differences: differences and oppositions which bear meaning. What differentiates a polite from an impolite greeting, a fashionable from an unfashionable garment? One comes to study not isolated signs but a system of distinctions.

THE DOMAIN OF SEMIOLOGY

Saussure's proposals concerning semiology were not immediately taken up, and it was only towards the middle of this century, many years after the publication of the *Course*, that others began to realize the importance of his suggestions. It is as if the individual disciplines had to develop in their own ways and rediscover Saussure's insights for themselves before they could become properly semiological. Indeed, what is now called 'structuralism' arose when anthropologists, literary critics, and others saw that the example of linguistics could help to justify what they sought to do in their own disciplines; and as they began to take linguistics as a model they realized that they were in fact developing the semiology which Saussure had so long ago proposed.

Thus, it was not until 1961, in his inaugural lecture at the Collège de France, that the anthropologist Claude Lévi-Strauss defined anthropology as a branch of semiology and paid homage to Saussure as the man who, in his discussion of semiology, had laid the foundations for the proper conception of anthropology. But fifteen years earlier, in an epoch-making article on 'Structural Analysis in Linguistics and Anthropology', Lévi-Strauss had already drawn upon the concepts and methods of linguistics to establish his brand of structuralism.

In this article Lévi-Strauss speaks of the advances in linguistics, especially in phonology, which have made it a scientific discipline and remarks that 'phonology cannot help but play the same renovating role for the social sciences that nuclear physics, for example, played for the exact sciences.' He proposes that the anthropologist follow the example of the linguist and reproduce in his own field something comparable to the 'phonological revolution'. Pho-

93

nology studies not isolated terms but relations between terms, systems of relation; and phonology passes from the study of phenomena which are consciously grasped or known by speakers of a language to their 'unconscious infrastructure'. It seeks to identify, that is to say, systems of relations which are known only subconsciously. What lesson can the anthropologist draw from this? He can take it, Lévi-Strauss says, as an example in method: in order to analyse signifying phenomena, in order to investigate actions or objects which bear meaning, he should postulate the existence of an underlying system of relations and try to see whether the meaning of individual elements or objects is not a result of their contrasts with other elements and objects in a system of relations of which members of a culture are not already aware.[1]

Indeed, Nikolai Trubetzkoy, in his seminal *Principles of Phonology* (1939), had already outlined the methodological implications of phonological theory for the social sciences and thus had advanced the semiology proposed by Saussure. Whereas the phonetician is concerned with the properties of actual speech sounds, the phonologist is interested in the differential features which are functional in a particular language; he asks what phonic differences are linked with differences of meaning, how the differential elements are related to one another and how they combine to form words or phrases. It is clear, Trubetzkoy continues, that these tasks cannot be accomplished by the methods of the natural sciences, which are concerned with the intrinsic properties of natural phenomena themselves and not with the differential features which are the bearers of social significance. In other words, in the natural sciences there is nothing corresponding to the distinction between *langue* and *parole*: there is no institution or conventional system to be studied. The social and human sciences, on the other hand, are concerned with the social use of material objects and must therefore distinguish between the objects themselves and the system of distinctive or differential features which give them meaning and value.

Attempts to describe such systems, Trubetzkoy argues, are closely analogous to work in phonology. The example he cites is the study of clothing, as it might be carried out by an anthropologist or a sociologist. Many features of physical garments themselves which would be of great importance to the wearer are of no interest to the anthropologist, who is concerned only with those features that carry a social significance. Thus, the length of skirts might carry a lot of significance in the social system of a culture, while the material from which they were made did not. Or again, if I were to wear a yellow suit rather than a grey suit, that might have considerable social meaning, but the fact that I have a strong preference for grey suits rather than brown suits, or a dislike for woollen materials, might be purely personal choices which carried no social significance. Just as the phonologist tries to determine which differences in sound bear meaning and which do not, so the anthropologist or sociologist studying clothing would be trying to isolate those features of garments which carried social significance. He attempts to reconstruct the system of relations and distinctions which members of a society have assimilated and which they display in taking certain garments as indicating a particular life-style, social role, or social attitude. He is, in short, interested in those features by which garments are made into signs.

Like the linguist, the anthropologist or sociologist is attempting to make explicit the implicit knowledge which enables people within a given society to communicate and understand one another's behaviour. The facts he is trying to explain are facts about people's implicit knowledge: that a particular action is regarded as polite while another is impolite; that a particular garment is appropriate in one situation but inappropriate in another. Where there is knowledge or mastery of any kind, there is a system to be explained. This is the fundamental principle which guides one's extrapolation from linguistics into other disciplines. If the meanings assigned to objects or actions by members of a culture are not purely random phenomena, then there must be a semiological system of distinctions, categories, and rules of combination which one might hope to describe.

One could thus assign to semiology a vast field of enquiry: if everything which has meaning within a culture is a sign and therefore an object of semiological investigation, semiology would come to include most disciplines of the humanities and the social sciences. Any domain of human activity, be it music, architecture, cooking, etiquette, advertising, fashion, literature, could be approached in semiological terms.

The immediate objection to an imperialistic semiology, which sought in this way to encompass so many other disciplines, might be that the signifying phenomena which one encounters in these various domains are not all alike. Even if most human objects and activities are signs, they are not signs of the same type. This is an important objection, and one of the major tasks of semiology is to distinguish between different types of signs, which may need to be studied in different ways.

Various typologies of signs have been proposed, but three fundamental classes of signs seem to stand out as requiring different approaches: the icon, the index, and the sign proper (sometimes misleadingly called 'symbol'). All signs consist of a signifier and a signified, a form and an associated meaning or meanings; but the relations between signifier and signified are different for each of these three types of sign. An *icon* involves actual resemblance between signifier and signified: a portrait signifies the person of whom it is a portrait less by an arbitrary convention than by resemblance. In an *index* the relation between signifier and signified is causal: smoke means fire because fire is generally the cause of smoke; clouds mean rain if they are the sort of clouds which produce rain; tracks are signs of the type of animal likely to have produced them. In the *sign proper*, however, the relation between signifier and signified is arbitrary and conventional: shaking hands conventionally signifies greeting; cheese is by convention an appropriate food with which to end a meal.

What are the implications of this three-way division for semiology? The main consequence is to make the sign proper the central object of semiology and to make the study of other signs a specialized and secondary activity. Study of the way in which a drawing or a photograph of a horse represents a horse might form part of semiology, but it seems more properly the concern of a philosophical theory of representation than of a linguistically based semiology. Semiology must identify and characterize iconic signs, but the study of icons is not likely to be one of its central activities.

Indices are, from the semiologist's point of view, more worrying. If he places them within his domain then he risks taking all human knowledge for his province, for any science which attempts to establish causal relations among phenomena could be seen as a study of indices and thus placed within semiology. Medicine, for example, tries to relate diseases to symptoms: to have discovered the symptoms of a disease is to have identified the signs which betray the presence of that disease and, reciprocally, to have learned what these symptoms are signs of. Meteorology attempts to construct a system in order to relate atmospheric conditions to their causes and consequences and thus to read them as signs: as signs of weather conditions. Economic prediction depends on a proper reading of economic signs; economics is the discipline which identifies these signs and enables one to read them. In short, a whole range of disciplines tries to decipher the natural or social world; the methods of these disciplines are different, and there is no reason to think that they would gain substantially by being brought under the banner of an imperialistic semiology.

Signs proper, where the relation between signifier and signified is arbitrary or conventional, are then the central domain of semiology. Indeed, they require semiological investigation if their mechanisms are to be understood. In the absence of a causal link between signifier and signified which would enable one to treat each sign individually, one must try to reconstruct the semiotic system, the system

97

of conventions, from which a whole group of signs derive. Precisely because the individual signs are unmotivated, one must attempt to reconstruct the system, which alone can explain them.

However, one cannot exclude indices altogether from the domain of semiology, for they form an interesting and important borderline case. The fact is that any index may be used as a conventional sign. Once the causal or indexical relationship between a signifier and a signified is recognized by a culture, the particular signifier becomes associated with its signified and can be used to evoke that meaning even in cases where the causal relation is absent. For example, once it is generally recognized that smoke means fire I can use smoke to signify fire. The smoke produced by a smoke machine may be used in a play to signify fire, even though the smoke is not in this case being caused by fire. The index is here being used as a conventional sign.

Many indices, of course, can be used as conventional signs in this theatrical way: if an actor is made up to look as if he had measles we read his spots as signifying measles in a conventional way and do not believe that the spots are in his case causally connected with measles. But there is a large set of conventionalized indices which are especially interesting to the semiologist because they come to constitute what one might call the conventional social mythology of a culture. What we call 'status-symbols' are perhaps the best example. As the name itself suggests, these are not just indices of status but symbols of status; though they have some causal or intrinsic relation to the status they signify, they have been promoted by the conventions of a society to the rank of symbol and carry more meaning than their causal or indexical nature would entail. Thus, a Rolls-Royce is certainly an index of wealth in that one must be wealthy to own one, but social convention has made it a symbol of wealth, a mythical object which signifies wealth more imperiously than other objects which might be equally expensive. Among the many objects which are indices of wealth in that they are all expensive, it has been singled out by social usage as a symbol of wealth. The semiologist who is studying social life as a system of signs will certainly want to include conventionalized indices of this kind within his domain.[2]

Moreover, there is another way in which indices enter the domain of the semiologist. Within particular sciences the meanings of indices change with the configurations of knowledge. Medical symptoms, for example, are read and interpreted differently from one era to another as knowledge advances. There are changes both in what are identified as symptoms and in the way symptoms are interpreted. It thus becomes possible for the semiologist to study the changes in medicine as an interpretive system, as a way of reading and identifying signs. He would be trying to discover the conventions which determine or make possible the medical discourse of a period and permit indices to be read. In this investigation the semiologist would be interested not in the symptoms or indices themselves, nor in the 'real' causal relation between index and meaning, but in the reading of indices within a system of conventions.

What then is the domain of semiology? How far does its empire extend? It will obviously have variable boundaries; there are many things which can be treated semiologically but which need not necessarily be studied in this way. In fact, to characterize the domain of semiology one must simply identify the different sorts of cases it can encounter.

I. At the heart of the semiological enterprise are systems of conventional signs used for direct communication. These include, first, the various codes used to convey messages which are composed in an existing natural language such as English. Morse code, semaphore codes, braille, and all the codes devised for secrecy can be used to convey an English message. Secondly, there is a whole series of specialized codes used to convey a particular type of information to groups who may not share the same natural language: chemical symbols, traffic signals and road signs, silver assay marks, mathematical symbols, the signs used in airports, trains, etc., and finally the recondite symbolisms of heraldic or alchemical codes.[3] All these cases involve conventional signs based on explicit codes: since they are designed for easy and unambiguous communication there is an explicit procedure for encoding and decoding, such as looking up the item in question in a code book. Such codes are pure examples of semiological systems, but precisely because they are so straightforward it is usually an easy matter to describe the principles on which they are constructed and so they often prove much less interesting to the semiologist than less explicit and more complicated systems which fall into our next category.

II. More complicated than explicit codes are systems where communication undoubtedly takes place but where the codes on which the communication depends are difficult to establish and highly ambiguous or open-ended. Such is the case, for example, with literature. To read and understand literature one requires more than a knowledge of the language in which it is written, but it is very difficult to establish precisely what supplementary knowledge is required for satisfactory interpretation of literary works. Certainly one is not dealing with the sort of codes for which keys or code books could be supplied. However, precisely because one is dealing with an extremely rich and complicated communicative system, the semiological study of literature and of other aesthetic codes (such as the codes of painting and music) can be extraordinarily interesting.

The reason for the evasive complexity of these codes is quite simple. Codes of the first type are designed to communicate directly and unambiguously messages and notions which are already known; the code simply provides an economical notation for notions which are already defined. But aesthetic expression aims to communicate notions, subtleties, complexities which have not yet been formulated, and therefore, as soon as an aesthetic code comes to be generally perceived as a code (as a way of expressing notions which have already been articulated) then works of art tend to move beyond this code. They question, parody, and generally undermine the code while exploring its possible mutations and extensions. One might even say that much of the interest of works of art lies in the ways in which they explore and modify the codes which they seem to be using; and this makes semiological investigation of these systems both highly relevant and extremely difficult.

III. The third sort of case which semiology must confront covers social practices which may not at first seem to involve communication but which are highly codified and certainly employ a whole series of distinctions in order to create meaning. Ritual and etiquette of various kinds and the systems of convention governing food and clothing are obviously semiological systems: to wear one set of clothes rather than another is certainly to communicate something, albeit indirectly. But one can go further and say that the buildings we inhabit, the objects we purchase, and the actions we perform are of interest to the semiologist because all the categories and operations through which they are invested with meaning are fundamentally semiological. This is not to say that purchasing a house, for example, is primarily or essentially a communicative action, but only that the differences between houses are invested with meaning by a semiological system and that in choosing one house rather than another one is dealing with the image projected by the particular house (a country cottage, a modern maisonette, a crumbling Victorian semi) One may, for purely practical reasons, choose to purchase a house whose image seems uncongenial, but one is nonetheless involved in a semiological system. The task of the semiologist in dealing with clothing, commercial objects, pastimes, and all these other social entities, is to make explicit the implicit meanings they seem to bear and to reconstruct the system of connotations on which these meanings are based.

IV. Finally we come to the cases which I initially set aside as involving indices rather than signs proper: the disciplines of the social and natural sciences which try to establish relations of cause and effect between phenomena and for which the meaning of an object or an action is likely to be its causal antecedent or consequence, its significance in a causal scheme. As I have already mentioned, though these disciplines are not in themselves semiological, that does not mean that they need escape the attention of the semiologist. The objects which these disciplines study are not signs proper, but they themselves, as disciplines, as 'languages' or systems of articulation, may be studied as semiotic systems.

This is obvious in the case of sciences which are now discredited, such as astrology. Since we do not believe in the causal relations which astrologists established between the movements of the planets and the events of people's lives, it is easy to consider astrology as a system of conventions. The semiologist studying astrology would ask what were the rules or conventions which astrologers employed in attributing meaning to the configurations of the heavens. What were the conventions which one had to accept to be an astrologer?

We would not hesitate to admit that we are here dealing with a system of signs which might be elucidated. But in fact, if we think about the matter we can see that our semiological analysis would not be fundamentally affected if future discoveries were to prove that everything the astrologers had said were true. The same set of rules would still underlie astrological discourse, whether the predictions they yield are true or false. And so we can extend the bounds of semiology somewhat further: semiology can study the conventions which govern the discourse and interpretations of any discipline. But notice what this involves. To the semiologist the truth or falsity of the propositions of a discipline will be irrelevant. If everything which botany now asserts were to be disproved, that would not affect a semiological analysis of the conventions of botany as a system of signs. Botany is not the sum of true statements about plants but a system of discourse. At any given period there are a great many things which could be truly said about plants which do not fall within the realm of botany (e.g. that roses are systematically cultivated and dandelions systematically uprooted), and the semiologist is interested in the conventions which exclude some statements from the realm of botany and permit others. Though some disciplines, such as medicine, meteorology, psycho-analysis, and astrology, might lend themselves more easily to a semiological analysis, in that they are more obviously concerned with the reading and interpretation of signs, in fact at this level any system of discourse can be studied semiologically since it is itself a system of signs.

SEMIOLOGICAL ANALYSIS

Linguistics has served as the model for semiology and, as Saussure suggested, has drawn attention to the conventional nature of signs and the differential nature of meaning. But it will perhaps be evident from the diversity of the sign systems I have mentioned that the concepts and techniques of linguistic analysis may be much better suited for the investigation of some systems than of others. In all cases the analyst distinguishes *langue* from *parole*, tries to go behind the actions or objects themselves to the system of rules and relations which enables them to have meaning. And in most cases he will be able to identify syntagmatic and paradigmatic relations: the relations between elements which can be combined to form higher-level units and relations between elements which can replace one another and which therefore contrast with one another to produce meaning. But in some systems the syntax is so weak as to make syntagmatic relations almost non-existent. Traffic signs, for example, generally do not involve the combination of more than one unit, or if they do (as in signs where the shape indicates the presence of a hazard and the device specifies the sort of hazard) the syntagmatic relation is very simple and uninteresting. Alternatively, in some systems the set of elementary paradigmatic oppositions is extremely limited. In Morse code, for example, there are only two oppositions: noise versus pause and short versus long. Other systems are semantically very weak. The abominations of Leviticus list the animals one is permitted and forbidden to eat. One can, with some ingenuity, reconstruct the system of rules which assign significance to particular animals, but this system only produces two meanings: clean and unclean (i.e. permitted and forbidden).

But for most systems there do seem to be syntagmatic relations, paradigmatic contrasts, and a variety of meanings which can be produced by various contrasts and relations. In the food system, for example, one defines on the syntagmatic axis the combinations of courses which can make up meals of various sorts; and each course or slot can be filled by one of a number of dishes which are in paradigmatic contrast with one another (one wouldn't combine roast beef and lamb chops in a single meal: they would be alternatives on any menu). These dishes which are alternatives to one another often bear different meanings in that they connote varying degrees of luxury, elegance, etc.

Many semiological systems are complicated, however, by the fact that they rest on other systems, particularly that of language, and thus become 'second order' systems. Literature is one such system: it has language as its basis and its supplementary conventions are conventions about special uses of language. Thus, to take a simple example, the rhetorical figures such as metaphor, metonymy, hyperbole, synecdoche can be seen as operations of a second-order literary code. When Shakespeare writes 'But thy eternal summer shall not fade', his words are signs which have a literal meaning in the linguistic code of English, but the rhetorical figure of metaphor is part of a second-order literary code which allows one to use the linguistic signs, *eternal summer*, to mean something like 'a full, languorous beauty which will always remain at its peak'. And, furthermore, there is a convention of love poetry making hyperbolic compliment of this kind, which draws upon meta-phors of nature and natural processes, an appropriate form of praise.

Now it is obvious that the system of literature – the knowledge one must acquire, over and above knowledge of the language, in order to read and interpret literary works – does not involve explicit codes like those of traffic signs or of etiquette. One can learn about various ways of interpreting figurative language, about the conventions governing different literary genres, about types of literary structure or organization. But literature continually under-mines, parodies, and escapes anything which threatens to become a rigid code or explicit rules for interpretation. Traffic signs do not violate the code of traffic signs, but literary works are continually violating codes. And this is because literature is fundamentally an exploration of the possibilities of experience, a questioning and deepening of the categories in and through which we ordinarily view ourselves and the world. Literary codes have an important role in that they make possible this questioning and deepening process, just as rules of etiquette make it possible to be impolite. But literary works never lie wholly within the codes that define them, and this is what makes the semiological investigation of literature such a tantalizing enterprise.[4]

In a series of unpublished reflections on medieval German legends, Saussure shows his interest in the semiology of literature and his awareness of some of the problems it poses. A legend, he writes, 'is composed of a series of symbols in a sense which remains to be defined.' These symbols, though more difficult to define than the units of a language, are doubtless governed by the same principles as other signs, and 'they all form part of semiology'.[5] In the case of literature, as in that of language and other semiotic systems, the fundamental problem is one of identity. One is not dealing with fixed signs such that a given form will always have the same meaning wherever it appears. On the contrary, the literary work is always drawing upon signs which exist prior to it, 'combining them and continually drawing from them new meaning'. Indeed, considering the problem of characters in his German legends, Saussure reaches the conclusion that one is confronted with a whole series of elements (proper names, attributes, relations with other characters, actions) and that what one speaks of as the character himself is nothing other than the creation of the reader, the result of drawing together and combining all the disparate elements which one encounters as one reads through the text.[6]

Saussure has here hit upon an important system of convention in literature. The production of characters is governed by a set of cultural models which enable us, for example, to infer motives from action or the qualities of a person from his appearance. And so if we say that in the course of a given novel or story a character changes, what we are saying is that, in terms of our literary models of character, two actions or attributes which are attached to a single character are in opposition, are incompatible: that according to our notions of character if someone first does X and later does Y we can only make sense of this by saying that the character himself changed.

ANAGRAMS AND LOGOCENTRISM

Saussure's remarks on the semiology of literature are sketchy though perceptive, but there was another, closely related enterprise to which he devoted much time in his later years and on which he left voluminous notes, though he never ventured to publish anything on the subject. He developed the theory that Latin poets deliberately concealed anagrams of proper names in their verses. He believed he had discovered a supplementary sign system, a special set of conventions for the production of meaning, and he filled many notebooks with remarks on the various types of anagrams he had discovered (letters dispersed through the text sometimes in their correct order, sometimes with a change of order, sometimes in pairs or triplets, etc.). Thus, in the 13 opening lines of Lucretius's *De Rerum Natura*, which are an invocation to Venus, Saussure discovered three anagrams of this goddess's Greek name: Aphrodite.

This example is quite typical: Saussure looked for anagrams of proper names which were of some relevance to the content of the verses, and he was interested in anagrams which were repeated throughout a text, not just in the occasional and possibly coincidental anagram. Certainly he amassed an impressive number of cases, but there were two things which worried him and which made him leave his speculations unpublished. First, the question of intention was a crucial one: if this were really a convention of Latin poetry, a convention which poets followed, then why were there no references to the practice or discussions of it in classical texts? And secondly, the advice he sought about the statistical probability of anagrams of the sort he had discovered was inconclusive. As he said in a letter, 'I remain perplexed on the most important point, that is to say what one should think about the reality or fantasy of the whole business.'[7]

But of course the important question is what are we to think of it? Was it, as one critic has suggested, 'la folie de Saussure', Saussure's bit of madness, or was it, as others have argued, a radical critique of language and in particular of the sign? Was Saussure obsessed by a chimerical problem, or was he trying to invent a new way of reading, freeing himself from the constraints of conventional linguistic codes and sign relations?

I think we can say quite frankly that Saussure's work on anagrams is not in itself a critique of the sign or an attempt to destroy convention so as to leave readers free to produce meaning according to their own devices. Saussure assumed that anagrams were governed by very strict supplementary conventions and certainly did not think of discovery of anagrams in a text as a form of self-expression or an escape from constraint. Morever, for Saussure the anagrams did not reveal a secret, subversive meaning; they simply provided other words, in fact proper names, which emphasized what the text was already discussing; they reinforced the meaning carried by other signs of the text rather than undermining these signs.

What then can we say of Saussure's theory? One might place it in a psychoanalytic perspective and say that he discovered a particular case of what can be called the 'insistence of the letter in the unconscious'. In reading over something one has written it is quite a familiar experience to discover that one has, without meaning to, repeated a word in two different senses or used similar sounding words in close proximity; and the explanation presumably is that a key word has lingered in the subconscious and helped to determine the choice of subsequent words. Psychoanalytic evidence, especially the examples in Freud's *Psychopathology of Everyday Life*, suggests the importance of purely verbal connections, connections of a punning and anagrammatic sort, in the workings of the unconscious. Thus one would have every reason to expect that poetic language, which is not governed by specific communicative ends and which thus gives greater scope to associative processes, would involve a certain amount of anagrammatic repetition.

If, as Saussure believed, the convincing cases of anagram involve repetition. then one can relate anagrams to other poetic processes: in Baudelaire's line 'Je sent*is* ma gorge *s*errée par la main *terri*ble de l'hystérie, the italicized sounds *i s terri* reproduce exactly the final word, *hystérie*. Presumably the poet wanted a rich echoing sound texture and happened to create an anagram. Consider this stanza from a sonnet by Gerard Manley Hopkins:

As kingfishers catch fire, dragonflies draw flame;
As tumbled over rim in roundy wells
Stones ring; like each tucked string tells, each hung bell's
Bow swung finds tongue to fling out broad its name.

We could find dispersed here the sounds of *flame* (1. 4 *fl*ing ... n*ame*), *Christ* (*k*, *r*, and *i* in 1.1'; *st* twice in 1.3), etc., and many other words, but these potential anagrams seem less

important than the echoes of '*king fish*ers *c*atch *fire*' and
'h*ung* . . . sw*ung* . . . to*ngue.*' Rhymes, alliteration, and asson-
ance are the elements of anagrams, and when they are
present it probably does not matter whether complete
anagrams form, since the effects of richness and emphasis
will be much the same in any case.

The reason why some people who have been studying
sign systems and the semiology of literature have been
particularly interested in Saussure's work on anagrams is
their desire to break out of what they call the 'logocentrism'
of Western culture and their belief that in looking for
anagrams Saussure was moving from sign to letter and thus
breaking away from logocentric conceptions of meaning.[8]
In brief, logocentrism involves the belief that sounds are
simply a representation of meanings which are present
in the consciousness of the speaker. The signifier is but a
temporary representation through which one moves to get
at the signified, which is what the speaker, in that revealing
English phrase, 'has in mind'. And the written word is an
even more derivative and imperfect form: it is the representation of a sound sequence which is itself a representation
of the thought. Interpretation, by this model, is a nostalgic
and retrospective process, an attempt to recover the concepts which were present to the consciousness of the
speaker or writer at the time of writing. And of course for
logocentrism, as indeed is the case with Saussure, the sign
is the fundamental unit; phonemes and letters are simply
convenient devices which in combination can be used to
represent the essence of the sign, which is the signified.

Though crudely stated, this is certainly the central
tradition of Western thought and many of Saussure's
pronouncements would place him squarely within it. The
reasons for trying to escape from it are essentially two, one
logical, the other moral and political. The moral and
political argument is that meaning should not be something
that we simply recover but something that we produce or
create; interpretation should transform the world, not
merely attempt to recover a past – especially since recovery

is, in any case, an impossible goal. No one can ever grasp exactly what another person might have had in mind, especially if the various distances which separate them are great; and therefore rather than guiltily attempt an impossible task one should welcome the necessity of creative interpretation and think of oneself as presented with a series of marks or traces which one can use to produce thought and meaning. The reality of signs is no longer to be located in the signified, which is intangible and irrecoverable, but in the signifier, and especially in the material traces of written language which one can actively interpret in a liberated way.

How does Saussure's work on anagrams relate to this argument? It is at best an ambiguous case. Certainly Saussure thought that his work had value only if he were in fact recovering what Latin poets had in mind; he certainly did not seek the exhilaration of creative interpretation. But the opponents of logocentrism could justifiably argue that Saussure experienced all the attractions of bizarre and creative interpretation, which explains his perseverance in his enterprise, and that the guilt and perplexity he experienced derived from his historical situation and prove what a very bad thing logocentrism is: trapped in a logocentric perspective, Saussure was unable to accept the true, liberated nature of what he was in fact doing and so not only perplexed himself with doubts but imposed such strict constraints on what he was doing (such as the decision to look only for anagrams of relevant proper names) that he could not find liberation in it.

The philosophical argument against logocentrism is very different, but here Saussure plays a similarly ambiguous role. He continually asserts the priority of spoken language to written language and sees writing, according to best logocentric tradition, as an imperfect and derivative representation. However, his fundamental principles seem to work against logocentrism. How is this so?

First, it is clear that for Saussure one does not start with a concept or mental essence of some sort, choose a phonetic sequence to represent it, and then move on to another autonomous concept for which one finds another phonetic expression. As our discussion in Chapter Two should have made clear, for Saussure both signifier and signified are form rather than content and they are first and foremost differential objects. Both signifiers and signifieds come into being only through the oppositions which articulate a domain, only through differences which form a system. 'In the linguistic system there are only differences with no positive terms.'

Thus, Saussure does not think of there being fully articulated concepts prior to the existence of a system of signifiers. Nor can he logically maintain that phonic expression itself is in any way essential to this system of differences. Sound is simply a way of manifesting the signifiers of a language, which are themselves defined in oppositional and combinatory terms without any reference to phonic material. So he ought not to assert, as he does, the priority of the spoken language. But his theory has another consequence, which is perhaps even more important. If, as Saussure writes, the most precise characteristic of every sign is that it differs from other signs, then every sign in some sense bears the traces of all the other signs; they are co-present with it as the entities which define it. This means that one should not think, as logocentrism would like to, of the presence in consciousness of a single autonomous signified. What is present is a network of differences. If I utter the word *brown* the 'concept' present in my mind (if indeed there is a concept present at all) is not some essence but a whole set of oppositions. Indeed, ultimately we could say that the whole notion of a linguistic system, the whole notion of *la langue* as Saussure defines it, is that of networks of differences at the level of both signifier and signified – networks which are already in place, already inscribed or written, as it were, in the mind of the subject. The act of uttering is simply a transitory and hence imperfect way of using one network of differences (those of the signifier) to produce a form which can be interpreted in terms of the other network of differences (those of the signified). The meaning of *brown* is not some essence which was in my mind at the moment of utterance but a space in this interpersonal network of differences (the semantic system of the language).

Attempts to challenge logocentrism involve a host of extremely complex problems and have so far appeared only in very abstruse discussions, of which the most intelligent are the writings of Jacques Derrida (see Bibliography). The remarks above simply give some indications of the lines of argument and attempt to demonstrate Saussure's seminal and ambiguous situation, as one who asserts logocentric positions very clearly but whose work acts in various ways to undercut those positions.

There are, I think, two aspects of Saussure's work which this problem leads one to stress. First, it may now be clearer why Saussure should have insisted on the psychological reality of *la langue*, which he treats as a social product that the individual passively assimilates. As I suggested earlier, the unconscious becomes a space of representation; the whole system is inscribed therein. And we can now see why this is important: what one 'has in mind' while speaking or writing is not a form and meaning conjured up for a fleeting instant but the whole system of a language, more permanently inscribed.

It is thus possible to emphasize, as Saussure himself often did, that meaning or the signified is not an entity so much as a bundle of differential values, a space in a system of differences. To give the meaning of a word or a sentence is to fill up this space with other signs and verbally to characterize some of the differences which define this space (thus, to give the meaning of *la langue* involves, among other things, defining the difference between *langue* and *parole*). And since signifieds are so intangible we might well feel justified in granting priority to the signifier, which can actually appear before us as a written word, promising meaning and provoking us to set off in pursuit of it. But if we do this we must remember that it is only the promise of determinable signifieds – meanings determined by convention – which makes a form a signifier.

The problem of logocentrism also makes one look again at Saussure's insistence on the social nature of language, on language as a collective institution, which the individual has assimilated but which fundamentally belongs to the world rather than to him, and which is always something other than himself. One might say that Saussure's theory illustrates the 'otherness of meaning'. What my words mean is the meaning they can have in this interpersonal system from which they emerge. The system is already in place, as the ground or condition of meaning, and to interpret signs is to read them in terms of the system.

This may go some way towards meeting the objection that Saussure is trapped in logocentrism, but it does not make interpretation the sort of liberated productive process which some theorists might wish it to be. Indeed, they would argue that my formulation has simply replaced the individual subject by a semiological system, making the system rather than the individual consciousness the source and guarantor of meaning. This is so, but the answer to this objection is that there can be no production of meaning without system. If one were able to escape from semiotic systems entirely, if one could free oneself from their constraints, then one would be free to assign meaning arbitrarily but meaning would not be *produced*. Moreover, the meanings assigned would have to come from somewhere and, encountering no resistance, they would generally be facile, uninteresting.

This last point is especially important since it bears on the nature and function of semiotic systems generally. The most interesting and complex interpretations arise in cases where there is on the one hand a semiotic system and, on the other, objects, actions, texts, which are difficult to interpret in terms of that system. They are ambiguous in terms of the system; they seem to escape it; they violate what one thinks to be its rules. But since we are governed by the human semiological imperative, *Try to make sense of things*, we struggle with the refractory or evasive object, deepening and extending our notions of significance, modifying and extending the rules of the system. We encounter here a point made earlier about the semiological system of literature: if there were a straightforward and explicit semiotic code which provided an automatic interpretation for every literary work, literature would be of much less interest, and the first thing authors would do is to violate or go beyond the rules of this code.

Interesting objects, actions, and texts seem partly to evade the semiotic systems to which they are related, but it is nevertheless crucial that they relate to a system; for if there were no conventions in whose terms we felt obliged to read them we might simply assign them meaning. And in simply assigning meaning we should have no other resources than ourselves, no other resources than all the notions we had already been living with. We should make no discoveries, either about ourselves or about the world. It is only when we find it hard to interpret an object but think that it belongs to a system which we do not fully grasp that we extend ourselves and discover new resources in the self as we rise to a difficulty and find ways of relating it to the relevant semiological system. Moreover, this process leads to an increase in self-awareness, to a better understanding of the codes and operations by which we create meanings.

CONCLUSIONS

'In the whole history of science,' wrote the philosopher Ernst Cassirer, 'there is perhaps no more fascinating chapter than the rise of the new science of linguistics. In its importance it may very well be compared to the new science of Galileo which in the seventeenth century changed our whole concept of the physical world.' Chapters Two and Three have outlined Ferdinand de Saussure's role in the rise of modern linguistics and have suggested why this is a fascinating episode in recent intellectual history. But Cassirer's bold comparison of modern linguistics with the new science of Galileo is more difficult to evalute. What does it mean and how could it be substantiated?

For Cassirer the crucial and revolutionary aspect of modern linguistics is Saussure's insistence on the primacy of relations and systems of relations. Here, in its fundamental concepts and methodological premises, Saussure's theory of language is an exceptionally clear expression of the formal strategies by which a whole series of disciplines, from physics to painting, transformed themselves in the late nineteenth and early twentieth centuries and became modern.

The strategy can be stated most simply as a shift in focus, from objects to relations. It is relationships that create and define objects, not the other way around. The philosopher of science, Alfred North Whitehead, offers a general statement of the problem:

The misconception which has haunted philosophic literature throughout the centuries is the notion of 'independent existence'. There is no such mode of existence; every entity is to be understood in terms of the way it is interwoven with the rest of the universe.

And in his book *Science and the Modern World* he shows that new discoveries in science produced so many complexities that a fundamental shift in perspective was necessary if the various disciplines were to come to terms with themselves and their objects. Physics discovered that it was exceedingly difficult to explain electricity and electromagnetic phenomena in terms of discrete units of matter and their movement. The solution seemed to be to reverse the problem: instead of taking matter as prime and trying to define the laws governing its behaviour, why not take energy itself, electrical energy, as prime and define matter in terms of electromagnetic forces. This change in perspective leads to the discovery of new scientific objects: an electron is not a positive entity in the old sense; it is a product of a field of force, a node in a system of relations, which, like a phoneme, does not exist independently of these relations.

What Whitehead calls the 'materialism' of the nineteenth century, the empiricism which grants ontological primacy to objects, gives way, he says, to a 'theory of relativity' in the broadest sense: a theory based on the primacy of relations. 'On the materialist theory,' Whitehead writes, 'there is material which endures. On the organic theory the only endurances are structures of activity.' Emphasis falls on the structures. 'The event is what it is by reason of the unification within itself of a multiplicity of relationships.' Outside these systems of relations, it is nothing.

Saussure, of course, states these themes clearly, not as aspects of some diffuse world view but as methodological postulates which are necessary if language is to be properly analysed. And alongside Saussure's affirmations we may place the unequivocal statement of the painter Georges Braque: 'I do not believe in things; I believe in relationships.' This is, perhaps, the true Modernist credo. What is Cubism if not an assertion of the primacy of relationships? In Cubist paintings objects lose their hitherto unquestioned primacy; they emerge with difficulty from the interaction of lines and planes; the three-dimensional space which supports ordinary objects is broken down in an attempt to represent a variety of perspectives and relations simultaneously. Or again, in Modernist literature one can observe the shift by which both poetry and the novel become less directly mimetic, less concerned with the representation of recognizable objects and scenes, and more interested in effects of juxtaposition, where relational values – relations between words or among various types of discourse – become the primary constituents of the work of art.

In various fields or disciplines shifts in technique have led to a concentration on systems of relation. This is the basis of Cassirer's bold claim: that for the thought of our century the world is no longer essentially a collection of independent entities, of autonomous objects, but a series of relational systems.

This move from object to structure is indeed a major shift in our conception of the world, but it is not clear how far the role of Galileo should fall to Saussure and Saussurian linguistics. From a historical point of view, his theory of language seems an exceptionally clear expression of a shift which was taking place simultaneously, if less explicitly, in a variety of fields: an expression or example more than a primary cause. Indeed, it seems likely that if Saussure is ever to be cast in the role of twentieth-century Galileo, his right to that position will depend on the discipline and mode of thought which he was actually instrumental in founding: semiology. To bring us to see social life and culture in general as a series of sign systems which a linguistic model can help us to analyse – this is the contribution which might eventually make him comparable to Galileo.

But of course it is too early to judge the real significance of Saussure in the intellectual history of our century, for work in the field of semiology has only recently begun, and it is not yet clear whether it will indeed become a dominant intellectual movement of our time. If it does become a major presence, a central discipline, this will be due to the efforts of many people besides Saussure; but his vision of a semiology which would encompass linguistics while taking it as a model has led others to give concrete expression to the semiological perspective: man is a creature who lives among signs and must try not only to grasp their meaning but especially to understand the conventions responsible for their meaning. It is Saussure who stands behind the claim, which many people would today espouse, that to study man is essentially to study the various systems by which he and his cultures organize and give meaning to the world.

Textual Note

References to Saussure's *Course in General Linguistics* are given in the text and use the following abbreviations:

Course = Ferdinand de Saussure, *Course in General Linguistics.* Translated by Wade Baskin. London: Peter Owen, 1960; Fontana, 1974.

Cours = Ferdinand de Saussure, *Cours de linguistique générale.* Edited by Tullio de Mauro. Paris: Payot, 1973. This is the standard edition. The pagination of the text is the same as in earlier Payot editions.

Engler = Ferdinand de Saussure, *Cours de linguistique générale.* Critical Edition by Rudolf Engler. Wiesbaden: Otto Harrassowitz, 1967-74. This edition prints the students' notes from which the *Course* was constructed. I cite it only when referring to those notes.

In quoting from the *Course* I give page references to both the French and English editions. All translations are my own.

I am indebted to Kate Patterson, Wlad Godzich, and especially to J. L. M. Trim for their comments on the manuscript.

Notes

1. THE MAN AND THE COURSE (pages 13-17)
   1. Letter of 4 January 1894, in 'Lettres de F. de Saussure à Antoine Meillet', *Cahiers Ferdinand de Saussure* 21 (1964), p. 95.

2. SAUSSURE'S THEORY OF LANGUAGE (pages 18-51)
   1. I use expressions such as 'Saussure wrote' purely for convenience. As was mentioned in Chapter One, very few passages of the *Course* were actually written by Saussure.
   2. An important exception, which Saussure discusses at length but which I here leave aside, is the phenomenon known as 'analogy', in which new forms are created on the analogy of existing forms. This is an important factor in linguistic change, but Saussure argues that it is fundamentally a synchronic phenomenon. For discussion see Chapter 3, p. 67.
   3. 'Notes inédites de F. de Saussure', *Cahiers Ferdinand de Saussure* 12 (1954), pp. 63 & 55-6.

3. THE PLACE OF SAUSSURE'S THEORIES (pages 52-89)
   1. *The Study of Language in England, 1780-1860*, Princeton, 1967, p. 127. This is an excellent discussion of the history of linguistics, with a wider scope than its title indicates.
   2. Michel Foucault, *The Order of Things*, London, 1970, p. 296.
   3. The *Mémoire* and other highly technical papers can be found in the *Recueil des publications scientifiques de F. de Saussure*, Geneva, 1922.
   4. *Current Issues in Linguistic Theory*, The Hague, 1964, p. 23. For further discussion of Chomsky's theories and his place in the history of linguistics, see John Lyons' *Chomsky* in Fontana's Modern Masters Series.
   5. However, Wallis Reid argues that Saussure's weakness is really a strength: 'The Saussurian Sign as a Control in Linguistic Analysis'. *Semiotexte* I, 2 (1974).
   6. For discussion of this question, see Wallis Reid, *op. cit.*

Notes

4. SEMIOLOGY: THE SAUSSURIAN LEGACY (pages 90-117)

1. Lévi-Strauss's essay can be found in *Structural Anthropology*, London, 1968. For a concise assessment of his work on signs see Edmund Leach's *Lévi-Strauss* in Fontana's Modern Masters series.

2. For this aspect of semiology see Roland Barthes, *Mythologies*, London, 1972, especially the important theoretical discussion in the final essay.

3. Many systems of this sort are discussed by Georges Mounin, *Introduction à la sémiologie*, Paris, 1970.

4. For the structuralist and semiological study of literature, see Jonathan Culler, *Structuralist Poetics: Structuralism, Linguistics, and the Study of Literature*, London and Ithaca, 1975.

5. Quoted in Jean Starobinski, *Les Mots sous les mots*, Paris, 1971, p. 15.

6. Quoted in D'Arco Silvio Avalle, 'La sémiologie de la narrativité chez Saussure', in *Essais de la théorie du texte*, ed. C. Bouazis, Paris, 1973, p. 33.

7. Quoted in Starobinski, p. 138. Starobinski publishes extensive extracts from Saussure's notebooks on anagrams.

8. For the problem of logocentrism and its relation to Saussure's theories see Jacques Derrida, *De la grammatologie*, Paris, 1967; Julia Kristeva, 'Pour une sémiologie des paragrammes', in *Semiotikè*, Paris, 1969; and the special issue of *Recherches/Semiotext*, 'Les Deux Saussures', (number 16, September 1974).

Chronology

1857	Birth of Ferdinand de Saussure in Geneva
1872	Writes an 'Essai sur les langues'
1874	Begins study of Sanskrit
1875-6	Studies Physics and Chemistry at the University of Geneva
1876	Joins the Société de linguistique de Paris
1876-8	Studies historical linguistics at the University of Leipzig
1878	*Mémoire sur le système primitif des voyelles dans les langues indo-européennes* is published
1878-9	Studies historical linguistics in Berlin
1880	Receives his doctorate *summa cum laude* from Leipzig for a thesis *De l'emploi du génitif absolu en sanscrit*
1880	Moves to Paris
1881-91	Maître de conférences at the École pratique des hautes études (teaching historical linguistics)
1891	Named Chevalier de la Légion d'honneur; becomes Professor at the University of Geneva
1907	First series of lectures on general linguistics
1908-9	Second series of lectures on general linguistics
1910-11	Third series of lectures on general linguistics
1913	Dies after several months' illness
1916	First edition of the *Cours de linguistique générale*, edited by Bally and Sechehaye

Bibliography

I. SAUSSURE'S WRITINGS

Course in General Linguistics, translated by Wade Baskin. London: Peter Owen, 1960; Fontana, 1974. (abbreviated as *Course*)

Cours de linguistique générale, edited by Tullio de Mauro. Paris: Payot, 1973. (abbreviated as *Cours*)

Cours de linguistique générale, critical edition by Rudolf Engler. Wiesbaden: O. Harrassowitz, 1967-74. Two volumes, four fascicules. (abbreviated as Engler)

The *Course* is essential reading, especially parts I, II, and III. Tullio de Mauro's admirable edition contains very full biographical and bibliographical information, explanatory comment, and quotations from the students' notes where they provide important variants. Engler's critical edition prints all the notes from which the *Course* was constructed.

II. ON SAUSSURE

Avalle, D'arco Silvio. 'La sémiologie de la narrativité chez Saussure', in *Essais de la théorie du texte,* ed. Charles Bouazis. Paris: Galilée, 1973.

Benveniste, Emile. 'Saussure après un demi-siècle', in *Problèmes de linguistique générale.* Paris: Gallimard, 1966. English translation = *Problems in General Linguistics.* Miami: University of Miami Press 1971.

Cahiers Ferdinand de Saussure. Geneva: 1941- .

Derossi, Giorgio. *Segno e struttura linguistici nel pensiero di F. de Saussure.* Udine: Del Bianco, 1965.

Derrida, Jacques. *De la grammatologie.* Paris: Minuit, 1967.

Godel, Robert. *Les Sources manuscrites du Cours de linguistique générale de F. de Saussur* . Geneva & Paris: Droz, 1957.

Koerner, E. F. K. *Bibliographia Saussuriana.* Metuchen, N.J.: Scarecrow Press 1972.

Koerner, E. F. K. *Ferdinand de Saussure: The Origin and Development of his Linguistic Thought in Western Studies of Language.* Braunschweig: Vieweg, 1973.

Bibliography

Starobinski, Jean. *Les Mots sous les mots: les anagrammes de F. de Saussure*. Paris: Gallimard, 1971.

'The Two Saussures'. Vol. I, *Semiotexte* I, 2 (Fall, 1974). Vol. II ('Saussure's Anagrams'), *Semiotexte* II, 1 (1975) = 'Les Deux Saussures', *Recherches* 16 (September 1974).

The most perceptive study of Saussure's thought is Derossi, unfortunately available only in Italian. Koerner's bibliography lists a vast number of writings on Saussure and related topics. His study of Saussure also provides a wealth of information, especially about verbal resemblances between Saussure and possible sources, but is less good on theoretical matters. Starobinski publishes many extracts from Saussure's work on anagrams and Avalle quotes and discusses his work on medieval German legends. *Cahiers Ferdinand de Saussure* is a periodical which publishes many important articles on Saussure. Derrida investigates philosophical problems in Saussure's thought. 'The Two Saussures' includes excellent discussions of linguistic theory and remarks on the anagrams. Godel's interesting study of manuscript sources is largely superseded by Engler's critical edition.

III. ON THE HISTORY OF LINGUISTICS

Aarsleff, Hans. *The Study of Language in England*, 1780-1860. Princeton: Princeton University Press, 1967.

Foucault, Michel. *Les Mots et les choses*. Paris: Gallimard, 1966. English translation = *The Order of Things*. London: Tavistock, 1970.

Lepschy, Giulio C. *A Survey of Structural Linguistics*. London: Faber, 1970.

Robins, R. H. *A Short History of Linguistics*. London: Longmans, 1967.

Foucault's discussion in chapters 4, 7, and 8 is extremely stimulating. For more sober surveys, see Lepschy for the modern period and Robins for the pre-modern. Aarsleff is especially good on the eighteenth century.

IV. SEMIOLOGY

Barthes, Roland. *Eléments de sémiologie*, in *Communications* 4 (1964). English translation = *Elements of Semiology*. London: Cape, 1967.

Culler, Jonathan. *Structuralist Poetics: Structuralism, Linguistics and the Study of Literature*. London: Routledge, and Ithaca: Cornell, 1975.

Derrida, Jacques. *Marges de la philosophie*. Paris: Minuit, 1972.

Guiraud, Pierre. *La Sémiologie*, Collection 'Que sais-je?'. Paris: PUF, 1971. English translation = *Semiology*. London: Routledge, 1975.

Semiotica. The Hague: Mouton, 1969-

Of the two introductory books, Barthes and Guiraud, Barthes is the earlier and the more interesting in its theoretical discussions. *Semiotica* is the journal of the International Association for Semiotic Studies and illustrates the range of semiological investigation. Several articles in Derrida's book are devoted to philosophical problems of semiology.